T0264496

FRACTIONAL CALCULUS WITH APPLICATIONS FOR NUCLEAR REACTOR DYNAMICS

FRACTIONAL CALCULUS WITH APPLICATIONS FOR NUCLEAR REACTOR DYNAMICS

Santanu Saha Ray

Department of Mathematics
National Institute of Technology Rourkela
Orissa, India

CRC Press
Taylor & Francis Group
Boca Raton London New York

CRC Press is an imprint of the
Taylor & Francis Group, an **informa** business

CRC Press
Taylor & Francis Group
6000 Broken Sound Parkway NW, Suite 300
Boca Raton, FL 33487-2742

First issued in paperback 2017

© 2016 by Taylor & Francis Group, LLC
CRC Press is an imprint of Taylor & Francis Group, an Informa business

No claim to original U.S. Government works

ISBN-13: 978-1-4987-2727-3 (hbk)
ISBN-13: 978-1-138-89323-8 (pbk)

Library of Congress Cataloging-in-Publication Data

Ray, S. (Santanu), author.
    Fractional calculus with applications for nuclear reactor dynamics / Santanu Saha Ray.
        pages cm
    Includes bibliographical references and index.
    ISBN 978-1-4987-2727-3 (hardcover : acid-free paper) 1. Nuclear
    reactors--Mathematical models. 2. Fractional calculus. I. Title.

QC786.R38 2015
621.48'30151583--dc23   2015026100

Visit the Taylor & Francis Web site at
http://www.taylorandfrancis.com

and the CRC Press Web site at
http://www.crcpress.com

Contents

Preface

The diffusion theory model of neutron transport plays a crucial role in reactor theory since it is simple enough to allow scientific insight and sufficiently realistic to enable study of many important design problems. Here, neutrons are characterized by a single energy or speed, and this phenomena allows for preliminary design estimates. The mathematical methods used to analyze such a model are the same as those applied in more sophisticated methods, such as multigroup diffusion theory and transport theory. The derivation of the neutron diffusion equation depends on Fick's law. The use of this law in reactor theory leads to the diffusion approximation, which is a result of a number of simplifying assumptions. On the other hand, higher-order neutron transport codes have been deployed in nuclear engineering for mostly time-independent problems and out-of-core shielding calculations. Given the limitations of the two-group diffusion theory, it is desirable to implement the more accurate, but also more expensive, multigroup transport approach.

The neutron diffusion and point kinetic equations are the most vital models of nuclear engineering, which are included in countless studies and applications in neutron dynamics. With the help of the neutron diffusion concept, we understand the complex behavior of average neutron motion. The simplest group diffusion problems involve only one group of neutrons, which for simplicity are assumed to be all thermal neutrons. A more accurate procedure, particularly for thermal reactors, requires splitting neutrons into two groups, in which case thermal neutrons are included in one group called the thermal or slow group and all the other neutrons are included in a fast group. The neutrons within each group are lumped together and their diffusion, scattering, absorption, and other interactions are described in terms of suitably average diffusion coefficients and cross sections, which are collectively known as group constants.

The variational iteration method and modified decomposition method have been applied to obtain the analytical approximate solution of the neutron diffusion equation with fixed source. Analytical methods like the homotopy analysis method and Adomian decomposition method have been used to obtain the analytical approximate solutions of the neutron diffusion equation for both finite cylinders and bare hemisphere. In addition to these methods, boundary conditions like zero flux as well as extrapolated boundary conditions are investigated. The explicit solution for critical radius and flux distributions is also calculated. The solution obtained is in the explicit form, which is suitable for computer programming and other purposes such as analysis of flux distribution in a square critical reactor. The homotopy analysis method is a very powerful and efficient technique that yields analytical solutions. With the help of this method, we can solve many functional equations, such as ordinary, partial differential, and integral equations. It does not require much memory space in a computer and is free from rounding-off errors and discretization of space variables. In this book, solutions obtained using these excellent methods are exhibited graphically.

Fractional differential equations appear more and more frequently in different research areas of applied science and engineering. Fractional neutron point kinetic equations have been solved using the explicit finite difference method, which is a very efficient and convenient numerical technique. The numerical solution for this fractional model is not only investigated but also presented graphically.

Fractional kinetic equations have proved to be particularly useful in the context of anomalous slow diffusion. The fractional neutron point kinetic model has been analyzed for the dynamic behavior of neutron motion. In this model, the relaxation time associated with variation in the neutron flux involves a fractional order acting as an exponent of the relaxation time. This approach is used to ensure optimal operation of a nuclear reactor. Results of dynamic behavior for neutron diffusion for subcritical, supercritical, and critical reactivities, as well as for different values of fractional order, are analyzed and compared with those of classical neutron point kinetic equations.

In the dynamical system of a nuclear reactor, point kinetic equations represent the coupled linear differential equations for neutron density and delayed neutron precursor concentrations. Point kinetic equations are essential in the field of nuclear engineering. With the help of the neutron diffusion concept, we understand the complex behavior of average neutron motion, which is diffused at very low or high neutron density. The modeling of these equations characterizes the time-dependent behavior of a nuclear reactor. The standard deterministic point kinetic model has been the subject of countless studies and applications to understand neutron dynamics and its effects. The reactivity function and neutron source term represent the parametric quantity of this neutron diffusion system. The dynamical process explained by the point kinetic equations is stochastic in nature. Neutron density and delayed neutron precursor concentrations differ randomly with respect to time. At high-power levels, random behavior is imperceptible. But at low-power levels, such as at the beginning of nuclear power generation, random fluctuation in neutron density and neutron precursor concentrations can be crucial. The numerical methods like the Euler–Maruyama method and strong order 1.5 Taylor method, which are used as powerful solvers for the stochastic neutron point kinetic equation, have been applied. The main advantage of the Euler–Maruyama method and strong order 1.5 Taylor method is that they do not require piecewise constant approximation over a partition for reactivity function and source functions.

The differential transform method (DTM) has been applied to compute the numerical solution for classical neutron point kinetic equations in nuclear reactors. DTM is an iterative procedure for obtaining analytic Taylor series solutions of differential equations. The new algorithm-based multistep

DTM (MDTM) is also applied here for solving the classical neutron point kinetic equation. The MDTM is treated as an algorithm in a sequence of intervals for finding a simple and accurate solution. Moreover, numerical examples with variable step, ramp, and sinusoidal reactivities are used to illustrate the preciseness and effectiveness of the proposed method.

The fractional stochastic neutron point kinetic equation is also solved using the explicit finite difference scheme. The numerical solution of the fractional stochastic neutron point kinetic equation has been obtained efficiently and elegantly. The fractional stochastic neutron point kinetic model has been analyzed for the dynamic behavior of the neutron. The explicit finite difference method has been implemented over the experimental data, with given initial conditions and step reactivities. The computational results indicate that this numerical approximation method is straightforward, effective, and easy for solving fractional stochastic point kinetic equations. Numerical results showing the behavior of neutron density and precursor concentration have been presented graphically for different values of fractional order. Most importantly, the random behavior of neutron density and neutron precursor concentrations have been analyzed in fractional order, something that has not been done before. The results of these numerical approximations for the solution of neutron population density and sum of precursor population for different arbitrary values of α have been computed numerically. In nuclear reactors, continuous indication of neutron density and its rate of change are important for safe start-up and operation.

The Haar wavelet operational method (HWOM) has been used to obtain the numerical approximate solution of the neutron point kinetic equation with time-dependent and -independent reactivity function in nuclear reactors. Using HWOM, stiff point kinetic equations have been analyzed elegantly with step, ramp, zig-zag, sinusoidal, and pulse reactivity insertions. On finding the solution for a stationary neutron transport equation in a homogeneous isotropic medium, the Haar wavelet collocation method has also been used. Recently, the Haar wavelet method has gained the reputation of being a very easy and effective tool for many practical applications in applied science and technology. To demonstrate the efficiency of the method, some test problems have been discussed.

Santanu Saha Ray

Acknowledgments

I take this opportunity to express my sincere gratitude to Dr. R. K. Bera, former professor and head, Department of Science, National Institute of Technical Teacher's Training and Research, Kolkata, India, and Dr. K. S. Chaudhuri, professor, Department of Mathematics, Jadavpur University, Kolkata, India, for their encouragement in the preparation of this book. I acknowledge the valuable suggestion rendered by scientist Shantanu Das and senior scientist B. B. Biswas, former head, Reactor Control Division, Bhaba Atomic Research Centre, Mumbai, India. It is not out of place to acknowledge the effort of my PhD scholar (Dr. A. Patra), who worked with me on my completed BRNS (under Department of Atomic Energy, Government of India) project entitled "Fractional Calculus Applied to Describe Reactor Kinetics and Flux Mapping." The research work related to this field has greatly inspired me to write this book.

I also express my sincere gratitude to the director of the National Institute of Technology, Rourkela, India, for his kind cooperation during the preparation of this book. I have received considerable assistance from my colleagues in the National Institute of Technology, Rourkela, India.

I wish to express my sincere thanks to those who were involved in the preparation of this book.

Moreover, I am especially grateful to the CRC Press/Taylor & Francis Group for their cooperation in all aspects of the production of this book.

I look forward to receiving comments and suggestions on the work from students, teachers, and researchers.

Author

Dr. S. Saha Ray is currently an associate professor at the Department of Mathematics, National Institute of Technology, Rourkela, India. Dr. Saha Ray completed his PhD in 2008 from Jadavpur University, Kolkata, India. He received a Master of Computer Applications degree in 2001 from the Indian Institute of Engineering Science and Technology, the then Bengal Engineering College, Sibpur, Howrah, India. He completed a master's degree in applied mathematics at the Calcutta University, Kolkata, India, in 1998 and a bachelor's (honors) degree in mathematics at St. Xavier's College, Kolkata, India, in 1996. Dr. Saha Ray has about 15 years of teaching experience at the undergraduate and postgraduate levels. He also has about 14 years of research experience in various fields of applied mathematics. He has published many research papers in numerous fields and various international SCI journals of repute, such as *Transaction of the ASME—Journal of Applied Mechanics, Annals of Nuclear Energy, Physica Scripta, Applied Mathematics and Computation,* and *Communications in Nonlinear Science and Numerical Simulation.* He has authored a book entitled *Graph Theory with Algorithms and Its Applications: In Applied Science and Technology* published by Springer. He has also published several articles on fractional calculus, mathematical modeling, mathematical physics, stochastic modeling, integral equations, and wavelet transforms. He is a member of the Society for Industrial and Applied Mathematics and the American Mathematical Society. He was the principal investigator of the BRNS research project, with grants from BARC, Mumbai, India. Currently, he is the principal investigator of a research project financed by the Department of Science & Technology, Government of India, and also the principal investigator of two research project financed by BRNS, BARC, Mumbai, India and NBHM, DAE, Government of India. He is supervising four research scholars, including two senior research fellows. One research fellow under his sole supervision has recently completed her Ph. D. in 2014 from the National Institute of Technology Rourkela, India. Dr. Saha Ray has been a lead guest editor for the international SCI journals of Hindawi Publishing Corporation, United States. Currently, he is the editor-in-chief for Springer's journal entitled *International Journal of Applied and Computational Mathematics.*

Preliminaries: Elements of Nuclear Reactor Theory

Nuclear Reactor Theory and Reactor Analysis

We begin with an overview of nuclear reactors and how nuclear energy is extracted from reactors. Here, nuclear energy means the energy released in nuclear fission. This energy release occurs because of the absorption of neutrons by fissile material. Neutrons are released by nuclear fission, and since the number of neutrons released is sufficiently greater than 1, a chain reaction of nuclear fission can be established. This allows, in turn, for energy to be extracted from the process. The amount of extracted energy can be adjusted by controlling the number of neutrons released. The higher the power density, the greater the economic efficiency of the reactor. The energy is extracted usually as heat via the coolant circulating in the reactor core. Finding the most efficient way to extract the energy is a critical issue. The higher the coolant output temperature, the greater the energy conversion efficiency of the reactor. Ultimately, consideration of material temperature limits and other constraints make power density levels uniform, which means careful control of neutron distribution. If there is an accident in a reactor system, the power output will run out of control. This situation is almost the same as an increase in the number of neutrons. Thus, the theory of nuclear reactors can be considered as the study of the behavior of neutrons in a nuclear reactor.

Here, *nuclear reactor theory* and *reactor analysis* are used to mean nearly the same thing. The term *reactor physics* is also sometimes used. This topic addresses neutron transport, including changes in neutron characteristics. Basically, a nuclear reactor is a device in which controlled nuclear fission chain reactions can be maintained. These nuclei fission into lighter nuclei (fission products) accompanied by the release of energy ($\cong 200$ MeV per event) plus several additional neutrons. Again these fission neutrons can then be utilized to induce still further fission reactions. The study of how to design a reactor so there is a balance between the production of neutrons in fission reactions and the loss of neutrons due to capture or leakage is known as nuclear reactor theory, nuclear reactor physics, or neutronics.

Discovery of the Neutron, Nuclear Fission, and Invention of the Nuclear Reactor

Technology generally progresses gradually by the accumulation of basic knowledge and technological developments. In contrast, nuclear engineering was born with the unexpected discovery of neutrons and nuclear fission, leading to a rather sudden development of the technology. The neutron was discovered by Chadwick in 1932. This particle had previously been observed by Irene and Frederic Joliot-Curie. However, they interpreted the particle as being a high energy γ-ray. The discovery of neutrons clarified the basic structure of the atomic nucleus (often referred to as simply the *nucleus*), which consists of protons and neutrons. Since the nucleus is very small, it is necessary to bring reacting nuclei close to each other in order to cause a nuclear reaction. Since nuclei have a positive charge, a very large amount of energy is required to bring nuclei close enough so that a reaction can take place. However, the neutron has no electric charge; thus, it can easily be brought close to a nucleus.

Nuclear fission was discovered by Hahn, Stresemann, and Meitner in 1939. Fission should have taken place in Fermi's experiments in 1939. The fact that Fermi did not notice this reaction indicates that nuclear fission is an unpredictable phenomenon. In 1942, Fermi created a critical pile after learning about nuclear fission and achieved a chain reaction of nuclear fission. The output power of the reaction was close to nil; however, this can be considered the first nuclear reactor made by a human being. However, this does not mean that a nuclear reactor can be built simply by causing fissions, by bombarding nuclei with neutrons. The following conditions have to be satisfied for nuclear fission reactions to occur:

- Exoergic reaction
- Sustainable as a chain reaction
- Controllable

The first nuclear reactor was built by Fermi under a plutonium production project for atomic bombs. In a nuclear reactor, radioactive material is rapidly formed. Therefore, nuclear reactors have the following unique and difficult issues, which need not be considered for power generators using other sources:

- Safety
- Waste
- Nuclear proliferation

The products of nuclear fission were the atomic bombs using enriched uranium and plutonium. One of the reasons for nuclear fission was to secure energy sources. After the war, the energy problem remained a big issue.

Thus, large-scale development of nuclear engineering was started in preparation for the estimated exhaustion of fossil fuels. Light-water reactors, which were put into practical use in nuclear submarines, were established in many countries. These reactors are not a solution to the energy problem, since they can utilize less than 1% of natural uranium. Fast reactors, on the other hand, can use almost 100% of natural uranium. Nuclear reactors comprise the following:

- *Fuel*: Any fissionable material.
- *Fuel element*: The smallest sealed unit of fuel.
- *Reactor core*: Total array of fuel, moderator, and control elements.
- *Moderator*: Material of low mass number that is inserted into the reactor to slow down or moderate neutrons via scattering collisions. Examples are light water, heavy water, graphite, and beryllium.
- *Coolant*: A fluid that circulates through the reactor removing fission heat. Examples are liquid coolants (water and sodium) and gaseous coolants ($^{4}_{2}He$ and CO_2).
- *Control elements*: Absorbing material inserted into the reactor to control core multiplication. Commonly absorbing materials include boron, cadmium, gadolinium, and hafnium.

Nuclear engineering is an excellent technology by which tremendous amounts of energy can be generated from a small amount of fuel. In addition to power generation, numerous applications are expected in the future. In addition to their use in energy generation, neutrons are expected to be widely used as a medium in nuclear reactions.

Constitution of the Atomic Nucleus

An atom is made of protons, neutrons, and electrons. Among these, the proton and neutron have approximately the same mass. However, the mass of the electron is only 0.05% that of these two particles. The proton has a positive charge and its absolute value is the same as the electric charge of one electron (the elementary electric charge). The proton and neutron are called nucleons and they constitute the nucleus. An atom is constituted of a nucleus and electrons that circle the nucleus due to Coulombic attraction. Species of atoms and nuclei are called elements and nuclides, respectively. An element is determined by its proton number (the number of protons). The proton number is generally called the atomic number and is denoted by Z. A nuclide is determined by both the proton number and the neutron number (the number of neutrons denoted by N). The sum of the proton number and neutron number, namely, the nucleon number, is called the mass number and is denoted by A ($A = Z + N$). Obviously, a nuclide can be determined by the atomic number and mass number.

In order to identify a nuclide, A and Z are usually added on the left side of the atomic symbol as superscript and subscript, respectively. For example, there are two representative nuclides for uranium, described as $^{235}_{92}U$ and $^{238}_{92}U$. The chemical properties of an atom are determined by its atomic number. These nuclides are called isotopic elements or isotopes. If the mass numbers of two elements are the same but their atomic numbers are different, they are called isobars. If the neutron numbers are the same, they are called isotones.

Nuclear Reaction

In chemical reactions, electrons are shared, lost, or gained during the formation of new compounds. In these processes, the nuclei just sit there and watch the show, passively sitting by and never changing their identities. In nuclear reactions, the roles of the subatomic particles are reversed. The electrons do not participate in the reactions; instead, they stay in their orbitals while the protons and neutrons undergo changes. Nuclear reactions are accompanied by energy changes that are a million times greater than those in chemical reactions. Energy changes are so great that changes in mass are detectable. Also, nuclear reaction yields and rates are not affected by the same factors (e.g., pressure, temperature, and catalysts) that influence chemical reactions. When nuclei are unstable, they are termed radioactive. The spontaneous change in the nucleus of an unstable atom that results in the emission of radiation is called radioactivity and this process of change is often referred to as the *decay* of atoms. The following lists salient features of nuclear reactions:

- Atoms of one element are typically converted into atoms of another element.
- Protons, neutrons, and other particles are involved; orbital electrons rarely take part.
- Reactions are accompanied by the relatively large changes in energy and measurable changes in mass.
- Reaction rates are affected by the number of nuclei, not by temperature, catalyst, or normally the compound in which an element occurs.

There are two types of nuclear reactions:

- *Fission*: The splitting of nuclei
- *Fusion*: The joining of nuclei (they fuse together)

Both reactions involve extremely large amounts of energy.

Nuclear fission is the splitting of one heavy nucleus into two or smaller nuclei, as well as some subatomic particles and energy. A heavy nucleus is usually unstable, due to the many positive protons pushing apart.

When fission occurs, energy is produced, more neutrons are given off, and neutrons are used to make nuclei unstable. It is much easier to crash a neutral neutron than a positive proton into a nucleus to release energy.

Delayed Neutrons

Delayed neutrons are the few neutrons (<1%) that appear with an appreciable time delay from the subsequent decay of radioactive fission products. These neutrons are vital for the effective control of the fission chain reaction.

Prompt Neutrons

Some of the fission neutrons appear essentially and instantaneously (within 10^{-14} s) of the fission event; these neutrons are called prompt neutrons.

Decay of a Nucleus

The decay of a nucleus is briefly explained in this section. Typical decays are α-, β-, and γ-decays, which result in emission of α-, β-, and γ-rays, respectively. An α-ray is the nucleus of ^{4}He, a β-ray is an electron, and a γ-ray is a high-energy photon. In β-decay, a positron may be emitted, which is called β+-decay. As a competitive process for γ-decay, internal conversion occurs when an orbital electron is ejected, rather than a γ-ray being emitted. Following α-decay, Z and N both decrease by 2. When a positively charged particle is emitted from a nucleus, the particle would normally have to overcome the potential of the Coulombic repulsive force, since the nucleus also has a positive charge. Spontaneous fission is an important feature of heavy nuclei. In this case, a Coulombic repulsive force even stronger than for α-decay applies, and the masses of the emitted particles are large; therefore, the parent nucleus must have sufficiently high energy. Since the neutron has no Coulomb barrier, a neutron can easily jump out of a nucleus if energy permits. Although it is not appropriate to call this decay, it is important in relation to the later-described delayed neutron emission, which accompanies nuclear fission.

Gamma Emission

Gamma emission involves the radiation of high energy or gamma (γ) photons being emitted from an excited nucleus. Those electrons cannot exist at such high energy levels indefinitely, the atom releases the energy absorbed, the electron falls, and the energy is released as a photon, which is of a specific energy—usually in the UV or visible region, but may also fall in the IR region. A nucleus that is excited will need to release its excess energy, and it does so by releasing a photon in the gamma region. The gamma photon is of much higher energy (shorter wavelength) than a UV- or visible-region photon. For example, when uranium-238 undergoes α-decay, a γ-ray is emitted.

$$^{238}_{92}\text{U} \rightarrow {}^{234}_{90}\text{Th} + {}^{4}_{2}\text{He} + {}^{0}_{0}\gamma$$

The time rate of change of the number of original nuclei present at that time is given by

$$-\frac{dN}{dt} = \lambda N(t)$$

where:
λ is the radioactive decay constant (s^{-1})

Here, $N(t) = N_0 e^{-\lambda t}$ where N_0 is the nuclei initially present. Radioactive half-life period is $N(T_{1/2}) = N_0/2 = N_0 e^{-\lambda(T_{1/2})}$ with $T_{1/2} = (\ln 2)/\lambda = 0.693/\lambda$.

β-Decay yields a daughter nucleus, which subsequently decays via delayed neutron emission as a delayed neutron precursor. There are six groups for delayed neutron precursors, characterized into six classes with half-life of approximately 55, 22, 6, 2, 0.5, and 0.2 s.

Cross Section

For a particular reaction, the probability of a neutron interacting with a nucleus is dependent upon not only the kind of nucleus involved but also the energy of the neutron. Accordingly, the absorption of a thermal neutron in most materials is much more probable than the absorption of a fast neutron. Also, the probability of interaction will vary depending upon the type of reaction involved. The probability of a particular reaction occurring between a neutron and a nucleus is called the microscopic cross section

(σ) of the nucleus for the particular reaction. This cross section will vary with the energy of the neutron. The microscopic cross section may also be regarded as the effective area the nucleus presents to the neutron for the particular reaction. The larger the effective area, the greater the probability for a reaction.

Because the microscopic cross section is an area, it is expressed in units of area, or square centimeters (cm^2). A square centimeter is tremendously large in comparison to the effective area of a nucleus; indeed, a physicist once referred to a square centimeter as being *as big as a barn* when applied to nuclear processes. The metaphor has persisted and microscopic cross sections are expressed in terms of barns. The relationship between barns and cm^2 is shown below.

$$1 \text{ barn} = 10^{-24} cm^2$$

Whether a neutron will interact with a certain volume of material depends not only on the microscopic cross section of the individual nuclei but also on the number of nuclei within that volume. Therefore, it is necessary to define another kind of cross section, known as the macroscopic cross section (Σ). The macroscopic cross section is the probability of a given reaction occurring per unit travel of the neutron. Σ is related to the microscopic cross section (σ) by the relationship shown below.

$$\Sigma = N\sigma$$

where:
  Σ is the macroscopic cross section (cm^{-1})
  N is the atom density of material (atoms/cm^3)
  σ is the microscopic cross section (cm^2)

The difference between the microscopic and macroscopic cross sections is that the microscopic cross section (σ) represents the effective target area that a single nucleus presents to a bombarding particle. The unit is barns or cm^2. The macroscopic cross section (Σ) represents the effective target area that is presented by all of the nuclei contained in 1 cm^3 of the material. The unit is 1/cm or cm^{-1}.

A neutron interacts with an atom of the material it enters in two basic ways. It interacts either through a scattering interaction or through an absorption reaction. The probability of a neutron being absorbed by a particular atom is the microscopic cross section for absorption, σ_a. The probability of a neutron scattering off of a particular nucleus is the microscopic cross section for scattering, σ_s. The sum of the microscopic cross section for absorption and the microscopic cross section for scattering is the total microscopic cross section, σ_t presented as

$$\sigma_t = \sigma_a + \sigma_s$$

Mean Free Path

If a neutron has a certain probability of undergoing a particular interaction in 1 cm of travel, then the inverse of this value describes how far the neutron will travel (in the average case) before undergoing an interaction. This average distance travelled by a neutron before interaction is known as the mean free path for that interaction and is represented by the symbol Λ. The relationship between the mean free path (Λ) and the macroscopic cross section (Σ) is $\Lambda = 1/\Sigma$.

Diffusion Coefficient

From diffusion theory, the diffusion coefficient is expressed in terms of macroscopic scattering cross section as

$$D = \frac{1}{3\Sigma_s}$$

where Σ_s is the macroscopic scattering cross section. In a weakly absorbing medium, where macroscopic absorption cross section is much less than macroscopic scattering cross section, that is, $\Sigma_a \ll \Sigma_s$, D becomes $D = 1/3\Sigma_{tr} = \lambda_{tr}/3$, where λ_{tr} is the transport mean free path (cm).

Neutron Flux

Consider the number of neutrons existing in 1 cm³ at any one instant and the total distance they travel each second while in that space. The number of neutrons existing in 1 cm³ of material at any instant is called neutron density and is represented by the symbol n with the unit of neutrons/cm³. The total distance these neutrons can travel each second will be determined by their velocity. A good way of defining neutron flux (ϕ) is to consider it to be the total path length covered by all neutrons in 1 cm³ in 1 s. Mathematically,

$$\phi = nv$$

where:
  ϕ is the neutron flux (neutrons/cm²-s)
  n is the neutron density (neutrons/cm³)
  v is the neutron velocity (cm/s)

Fick's Law and Neutron Diffusion Equation

Diffusion is a process that occurs when the concentration of a solute in one region of the solution is greater than that in another region of the solution; the solute diffuses from the region of higher concentration to the region of lower concentration.

Fick's law states that the current density vector is proportional to the negative gradient of the flux. Thus, Fick's law for neutron diffusion is given by

$$J = -DV\phi$$

where:
   J is the neutron current density, which is the net amount of neutrons that pass per unit time through unit area
   D is the diffusion coefficient
   ϕ is the neutron flux
   V is the del or gradient operator

The use of this law in reactor theory leads to the diffusion approximation. Diffusion theory is based on Fick's law and the continuity equation. To derive the neutron diffusion equation, we make the following assumptions:

1. We use a one-speed or one-group approximation where the neutrons can be characterized by a single average kinetic energy.

2. We characterize neutron distribution in the reactor by the particle density $n(t)$, which is the number of neutrons per unit volume at a position $\vec{r}$ at a time t. Its relationship to the flux is

$$\varphi(\vec{r},t) = v\,n(t)$$

We consider an arbitrary volume V and write the balance equation or equation of continuity thus:

$$\begin{bmatrix} \text{Time rate of} \\ \text{the number of} \\ \text{neutrons in } V \end{bmatrix} = \begin{bmatrix} \text{Production} \\ \text{rate in } V \end{bmatrix} - \begin{bmatrix} \text{Absorptions} \\ \text{in } V \end{bmatrix} - \begin{bmatrix} \text{Net leakage from} \\ \text{the surface of } V \end{bmatrix}$$

The first term is expressed mathematically as

$$\frac{d}{dt}\left[\int_V n(\vec{r},t)dV \right] = \frac{d}{dt}\left[\int_V \frac{1}{v}\varphi(\vec{r},t)dV \right] = \frac{1}{v}\frac{d}{dt}\left[\int_V \varphi(\vec{r},t)dV \right]$$

The production rate can be written as

$$\int_V S(\vec{r},t)dV$$

The absorption term is

$$\int_V \sum_a (\vec{r})\varphi(\vec{r},t)dV$$

and the leakage term is

$$\int_V \nabla \cdot J(\vec{r},t)dV$$

Here we have converted the surface integral to a volume integral by use of Gauss theorem or the divergence theorem.

Substituting for the different terms in the balance equation we get

$$\frac{1}{v}\frac{d}{dt}\left[\int_V \varphi(\vec{r},t)dV\right] = \int_V S(\vec{r},t)dV - \int_V \sum_a (\vec{r})\varphi(\vec{r},t)dV - \int_V \nabla \cdot J(\vec{r},t)dV$$

or we can obtain

$$\int_V \left[\frac{1}{v}\frac{\partial\phi(\vec{r},t)}{\partial t} - S + \sum_a \varphi + \nabla \cdot J(\vec{r},t)\right]dV = 0$$

Since the volume V is arbitrary we can write

$$\frac{1}{v}\frac{\partial\phi(\vec{r},t)}{\partial t} = -\nabla \cdot J(\vec{r},t) - \sum_a \varphi + S$$

We now use the relationship between J and φ (Fick's law) to write the following diffusion equation:

$$\frac{1}{v}\frac{\partial\phi(\vec{r},t)}{\partial t} = \nabla \cdot [D(\vec{r})\nabla\varphi(\vec{r},t)] - \sum_a (\vec{r})\varphi(\vec{r},t) + S(\vec{r},t)$$

This equation is the basis of much of the development in reactor theory using diffusion theory.

Reproduction Factor

Most of the neutrons absorbed in the fuel cause fission, but some do not. The reproduction factor (η) is defined as the ratio of the number of fast neutrons produced by thermal fission to the number of thermal neutrons absorbed in the fuel. The reproduction factor is shown below.

$$\eta = \frac{\text{Number of fast neutrons produced by thermal fission}}{\text{Number of thermal neutrons absorbed in the fuel}}$$

The reproduction factor can also be stated as a ratio of rates as

$$\eta = \frac{\text{Rate of production of fast neutrons by thermal fission}}{\text{Rate of absorption of thermal neutrons by the fuel}}$$

Effective Multiplication Factor

The infinite multiplication factor can fully represent only a reactor that is infinitely large, because it assumes that no neutrons leak out of the reactor. To completely describe the neutron life cycle in a real, finite reactor, it is necessary to account for neutrons that leak out. The multiplication factor that takes leakage into account is the effective multiplication factor (k_{eff}), which is defined as the ratio of the neutrons produced by fission in one generation to the number of neutrons lost through absorption and leakage in the preceding generation. The effective multiplication factor may be expressed mathematically as

$$k_{eff} = \frac{\text{Neutron production from fission in one generation}}{\text{Neutron absorption and neutron leakage in preceeding generation}}$$

The condition where the neutron chain reaction is self-sustaining and the neutron population is neither increasing nor decreasing is referred to as the critical condition, which is expressed by the simple equation $k_{eff} = 1$. If neutron production is greater than absorption and leakage, the reactor is said to be supercritical. In a supercritical reactor, $k_{eff} > 1$, and the neutron flux increases with each generation. If, on the other hand, the neutron production is less than the absorption and leakage, the reactor is said to be subcritical. In a subcritical reactor, $k_{eff} < 1$; this means that the flux decreases with each generation.

Buckling

Buckling is a measure of the extent to which the flux curves or *buckles*. In a nuclear reactor, criticality is achieved when the rate of neutron production is equal to the rate of neutron losses, including both neutron absorption and neutron leakage. Geometric buckling is a measure of neutron leakage, while material buckling is a measure of neutron production minus absorption. Thus, in the simplest case of a bare, homogeneous, steady-state reactor, the geometric and material buckling must be equal.

The diffusion equation is usually written as

$$\frac{\partial\phi(\vec{r},t)}{\partial t} = \nabla.\left[D(\phi,\vec{r})\nabla\phi(\vec{r},t)\right]$$

where:

$\phi(\vec{r},t)$ is the density of the diffusing material at location $\vec{r}$ and time t

$D(\phi,\vec{r})$ is the collective diffusion coefficient for density ϕ at location $\vec{r}$

∇ represents the vector differential operator del

If the diffusion coefficient depends on density, then the equation is nonlinear; otherwise it is linear.

If flux is not a function of time, then the buckling terms can be derived from the following diffusion equation:

$$-D\nabla^2\phi + \Sigma_a\phi = \frac{1}{k}\nu\Sigma_f\phi$$

where:

k is the criticality eigenvalue

ν is the number of neutrons emitted per fission

Σ_f is the macroscopic fission cross section

D is the diffusion coefficient

Rearranging the terms, the diffusion equation becomes

$$-\frac{\nabla^2\phi}{\phi} = \frac{\left(k_\infty/k\right)-1}{L^2} = B_g^{\,2}$$

where the diffusion length $L \equiv \sqrt{D/\Sigma_a}$ and $k_\infty = \nu\Sigma_f/\Sigma_a$.

We also can write

$$B_g^{\,2} = \frac{\left(k_\infty/k\right)-1}{L^2} = \frac{\left(1/k\right)\nu\Sigma_f - \Sigma_a}{D}$$

By diffusion theory,

$$k = k_{eff} = \frac{\nu\Sigma_f}{\Sigma_a + DB_g^2} \text{ or } k = \frac{\nu\Sigma_f/\Sigma_a}{1 + L^2 B_g^2}$$

Assuming the reactor is in a critical state ($k = 1$), geometric buckling is $B_g^2 = (\nu\Sigma_f - \Sigma_a)/D$.

By equating geometric and material buckling, one can determine the critical dimensions of a one region of a nuclear reactor.

- *One-group diffusion model*: All the neutrons of the reaction are characterized by a single kinetic energy of range from 10^{-3} to 10^7 eV.
- *Multigroup diffusion model*: The multigroup diffusion mode is a sophisticated model of neutron density behavior based on breaking up the range of neutron energies into intervals or *groups* and then describing the diffusion of neutrons in each of these groups separately, accounting for the transfer of neutrons between groups caused by scattering.

Reactivity

Reactivity is a measure of the departure of a reactor from criticality. Reactivity is related to the value of k_{eff}. Reactivity is a useful concept to predict how the neutron population of a reactor will change over time.

$$\rho = \frac{k_{eff} - 1}{k_{eff}}$$

Here, ρ may be positive, zero, or negative, depending upon the value of k_{eff}. The larger the absolute value of reactivity in the reactor core, the further the reactor is from criticality. It may be convenient to think of reactivity as a measure of a reactor's departure from criticality. Such processes, whereby the reactor operating conditions will affect the criticality of the core, are known as feedback reactivity. In reactivities, step, ramp, zig-zag, sinusoidal, pulse, and temperature feedback reactivity are used to solve neutron kinetic problems.

Mean Generation Time

The mean generation time, Λ, is the average time between the emission of a neutron and the capture of the neutron that results in fission. The mean generation time is different from the prompt neutron lifetime because the mean generation time includes only neutron absorptions that lead to fission reactions and not other absorption reactions. The two times are related by the following equation:

$$\Lambda = \frac{l}{k_{\text{eff}}}$$

Here, k_{eff} is the effective neutron multiplication factor.

Nuclear Reactor Dynamics

Nuclear reactor dynamics is the study of the time dependence of the related process in determining core multiplication as a function of the power level of the reactor.

Point Reactor Kinetic Model

The one-group diffusion model we have been using to study reactor criticality is also capable of describing the time behavior of a nuclear reactor, providing the effects of delayed neutrons. The model does not really treat the reactor as a point; it merely assumes that the flux shape does not change with time.

Delayed neutrons are extremely important for reactor time behavior. For thermal reactors and for fast reactors, prompt neutron lifetimes are 10^{-4} and 10^{-7} s, respectively. The reactor period predicted by this model is too small for effective control of reactor.

The importance of the delayed neutron on the reactor, the effective lifetime of such neutrons given by their prompt lifetime plus additional delay time λ_i^{-1} characterizing the β-decay of their precursor is considerably longer than prompt $l \approx 10^{-6}$ s–10^{-4} s. Hence, delayed neutrons substantially increase the time constant of a reactor so that effective control is possible. The neutron point kinetic model defined as

$$\frac{dn(t)}{dt} = \left[\frac{k(1-\beta)-1}{l}\right]n(t) + \sum_i \lambda_i c_i$$

$$\frac{dc_i}{dt} = \beta_i \left(\frac{k}{l}\right)n(t) - \lambda_i c_i$$

where:

λ_i is decay constant of the ith precursor group

β_i is the fraction of all fission neutrons emitted per fission that appear from the ith precursor group

$\beta = \sum_i \beta_i$ is total fraction of fission neutrons that are delayed

Nuclear energy is the only available energy source that could fulfill our current and future energy demands without polluting the earth any further. Preceding its useful utilization, such as generation of electricity, materials irradiation, or medical applications, its extraction within a nuclear reactor must occur. Design of nuclear reactors and analysis of their various operational modes is therefore a complicated task that encompasses several areas of science and engineering. At its start, however, determination of neutronic conditions within the reactor core plays a crucial role and has accordingly received substantial attention in the field of reactor physics over the past decades. The main objective of such neutronic analyses is to describe and predict the states of a reactor under various conditions and to determine the optimal configuration in which it is capable of long-term, self-sustained operation with minimal human intervention.

In the neutron diffusion theory, equations that govern the dynamics of space–time and neutron population are called kinetics equations. Kinetics equations are divided into point kinetic equations and space kinetics equations. In this work, I will emphasize the point kinetic model, more specifically, variations in the neutron density for small timescales, or equivalent changes in criticality due to changes of nuclear parameters in small time intervals. Point kinetic equations describe the variation of neutron density over time, assuming total separability of time from spatial degrees of freedom but with an a priori known spatial shape of the density. The point kinetic model plays a significant role in reactor physics and is used to estimate the power response of a reactor, allowing for control of and intervention in power plant operation that may help avoid the occurrence of incidents or accidents. The more recent approach is a fractional kinetics model, which reproduces the classical model and thus allows for capturing effects that differ from the usually employed hypothesis of Fick. The fractional point kinetic model presented here is derived thoroughly and solved numerically, which hopefully will mark the beginning of extensive research for future validation and applications of this approach in nuclear reactor theory. The diffusion theory model of neutron transport has played a crucial role in reactor theory. The neutron transport equation models the transport of neutral particles in scattering, fission, and absorption events with no self-interactions. Hence, the point kinetic model is widely used in reactor dynamics because of the apparent simplicity of the resulting equations.

1

Mathematical Methods in Nuclear Reactor Physics

1.1 Analytical Methods and Numerical Techniques for Solving Deterministic Neutron Diffusion and Kinetic Models

This chapter provides a brief description of the analytical and numerical methods to obtain the solution for neutron point kinetic models in the nuclear dynamical system. The basic ideas of linear and nonlinear analytical methods have been introduced to solve the kinetic equation in the field of nuclear reactor dynamics using the homotopy analysis method (HAM), variational iteration method (VIM), and Adomian decomposition method (ADM). Numerical techniques like the differential transform method (DTM), multistep differential transform method (MDTM), and explicit finite difference method (EFDM) have also been applied to obtain the numerical approximate solution for the neutron point kinetic equation in the nuclear reactor dynamical system.

1.1.1 Homotopy Analysis Method

Nonlinear equations are difficult to solve, especially fractional differential equations, partial differential equations, differential–integral equations, differential-difference equations, and coupled equations. Unlike perturbation methods, the HAM is independent of small/large physical parameters and provides us a simple way to ensure the convergence of a solution series. The method was first devised by Shijun Liao in 1992; since then many researchers have successfully applied this method to various nonlinear problems in science and engineering.

We consider the following differential equation:

$$N[u(x,t)] = 0 \tag{1.1}$$

where:
N is the nonlinear operator
x and t are the independent variables
$u(x,t)$ is an unknown function

For simplicity, we ignore all boundary or initial conditions that are treated in the same way. By means of generalizing HAM [1,2], we first construct the zeroth-order deformation equation.

$$(1-q)L\big[\phi(x,t;q)-u_0(x,t)\big]=qhH(x,t)N\big[\phi(x,t;q)\big] \qquad (1.2)$$

where:
   $q \in [0,1]$ is the embedding parameter
   $h \neq 0$ is an auxiliary parameter
   L is an auxiliary linear operator
   $\phi(x,t;q)$ is an unknown function
   $u_0(x,t)$ is an initial guess of $u(x,t)$
   $H(x,t)$ is a nonzero auxiliary function

For $q = 0$ and $q = 1$, the zeroth-order deformation equation given by Equation 1.2 leads to

$$\phi(x,t;0) = u_0(x,t) \quad \text{and} \quad \phi(x,t;1) = u(x,t) \qquad (1.3)$$

When the value of q increases from 0 to 1, the solution $\phi(x,t;q)$ varies from the initial guess $u_0(x,t)$ to the solution $u(x,t)$. Expanding $\phi(x,t;q)$ in Taylor's series with respect to q, we have

$$\phi(x,t;q) = u_0(x,t) + \sum_{m=1}^{\infty} u_m(x,t)q^m \qquad (1.4)$$

where:

$$u_m(x,t) = \frac{1}{m!}\frac{\partial^m \phi(x,t;q)}{\partial q^m}\bigg|_{q=0} \qquad (1.5)$$

The convergence of the series (Equation 1.4) depends upon the auxiliary parameter h. If it is convergent at $q = 1$, we get

$$u(x,t) = u_0(x,t) + \sum_{m=1}^{\infty} u_m(x,t) \qquad (1.6)$$

which must be one of the solutions of the original differential equation. Now we define the vector

$$\vec{u}_m(x,t) = \big\{u_0(x,t),\, u_1(x,t),\, u_2(x,t),\cdots,u_m(x,t)\big\} \qquad (1.7)$$

Differentiating the zeroth-order deformation equation (Equation 1.2) m times with respect to q, dividing them by $m!$, and finally setting $q = 0$, we obtain the following mth-order deformation equation:

$$L\big[u_m(x,t)-\chi_m u_{m-1}(x,t)\big]=hH(x,t)\Re_m\big[\vec{u}_{m-1}(x,t)\big] \qquad (1.8)$$

where:

$$\Re_m\left(\vec{u}_{m-1}\right) = \frac{1}{(m-1)!}\frac{\partial^{m-1}N\left[\phi(x,t;q)\right]}{\partial q^{m-1}}\Bigg|_{q=0} \tag{1.9}$$

and

$$\chi_m = \begin{cases} 0, & m \leq 1 \\ 1, & m > 1 \end{cases} \tag{1.10}$$

Now, by applying L^{-1} on both sides of mth-order deformation equation (Equation 1.8), we obtain the solution

$$u_m(x,t) = \chi_m u_{m-1}(x,t) + L^{-1}\left\{hH(x,t)\Re_m\left[u_{m-1}(x,t)\right]\right\} \tag{1.11}$$

In this way, it is easy to obtain u_m for $m \geq 1$; at the Mth order, we have

$$u(x,t) = \sum_{m=0}^{M} u_m(x,t) \tag{1.12}$$

When $M \to +\infty$, we obtain an accurate approximation of the original Equation 1.1.

It provides a simple way to ensure the *convergence* of the solution, freedom to choose the *basis functions* of the desired solution, and flexibility in determining the *linear operator* of the homotopy.

1.1.2 Adomian Decomposition Method

The ADM is a semi-analytical method for solving ordinary and partial nonlinear differential equations. The method was developed between the 1970s and the 1990s by George Adomian. The main aim of this method has been superseded by the more general theory of the HAM. The vital aspect of the method is employment of the *Adomian polynomials*, which allow for solution convergence of the nonlinear portion of the equation without simply linearizing the system. This algorithm provides the solution in a rapidly convergent series.

Let us consider the general form of a differential equation [3]:

$$Fy = g \tag{1.13}$$

where:
F is the nonlinear differential operator with linear and nonlinear terms
y is the unknown function and g is the known function

The differential operator is decomposed as

$$F \equiv L + R \tag{1.14}$$

where:
  L is an easily invertible linear operator
  R is the remainder of the linear operator

For our convenience L is taken as the highest-order derivative. Equation 1.13 can now be written as

$$Ly + Ry + Ny = g \tag{1.15}$$

where:
  Ny corresponds to the nonlinear term

Solving for Ly from the above equation, we have

$$Ly = g - Ry - Ny \tag{1.16}$$

Because L is invertible, L^{-1} is the integral operator.

$$L^{-1}(Ly) = L^{-1}(g) - L^{-1}(Ry) - L^{-1}(Ny) \tag{1.17}$$

If L is a second-order operator, L^{-1} is a twofold integral operator.

$$L^{-1} \equiv \int_0^t \int_0^t (\cdot) \, dt \, dt \quad \text{and} \quad L^{-1}(Ly) = y(t) - y(0) - ty'(0)$$

Now, solving Equation 1.17 for y yields

$$y = y(0) + ty'(0) + L^{-1}(g) - L^{-1}(Ry) - L^{-1}(Ny) \tag{1.18}$$

Let us consider the unknown function $y(t)$ in the infinite series as

$$y(t) = \sum_{n=0}^{\infty} y_n \tag{1.19}$$

The nonlinear term $N(y)$ will be decomposed by the infinite series of Adomian polynomials A_n ($n \geq 0$) [3,4] as

$$Ny = \sum_{n=0}^{\infty} A_n \tag{1.20}$$

where A_n is obtained by

$$A_n = \frac{1}{n!}\left[\frac{d^n}{d\lambda^n}N\left(\sum_{i=0}^{\infty}y_i\lambda^i\right)\right]_{\lambda=0} \tag{1.21}$$

Now, substituting Equations 1.19 and 1.20 into Equation 1.18, we obtain

$$\sum_{n=0}^{\infty}y_n = y(0)+ty'(0)+L^{-1}(g)-L^{-1}\left[R\left(\sum_{n=0}^{\infty}y_n\right)\right]-L^{-1}\left[\left(\sum_{n=0}^{\infty}A_n\right)\right] \tag{1.22}$$

Consequently we obtain

$$y_0 = y(0)+ty'(0)+L^{-1}(g)$$
$$y_1 = -L^{-1}R(y_0)-L^{-1}(A_0)$$
$$y_2 = -L^{-1}R(y_1)-L^{-1}(A_1) \tag{1.23}$$
$$\dots$$
$$y_{n+1} = -L^{-1}R(y_n)-L^{-1}(A_n)$$

and so on.

Based on the ADM, we shall consider the solution $y(t)$ as

$$y \cong \sum_{k=0}^{n-1}y_k = \varphi_n \quad \text{with} \quad \lim_{n\to\infty}\varphi_n = y(t)$$

We can apply this method to many real physical problems, and the obtained results have proved to be of high accuracy. In most of the problems, the practical solution φ_n, the n-term approximation, is convergent and accurate even for small values of n.

1.1.3 Modified Decomposition Method

A powerful modification of the ADM [3,4] has been proposed by noted researcher Wazwaz [5]. A large amount of literature has developed concerning the ADM and the related modifications to investigate various scientific models.

We consider the following general differential equation:

$$Lu + Ru + Nu = g(t) \tag{1.24}$$

where:
  L is the operator of the highest-order derivative with respect to t
  R is the remainder of the linear operator
  Nu is the nonlinear term

Thus, we obtain

$$Lu = g(t) - Ru - Nu \qquad (1.25)$$

Since L is an easily invertible linear operator, the inverse L^{-1} is assumed to be an integral operator given by

$$L_t^{-1} = \int_0^t (\cdot) dt \qquad (1.26)$$

Operating the integral operator L^{-1} on both sides of Equation 1.25, we get

$$u = \varphi + L^{-1}g(t) - L^{-1}Ru - L^{-1}Nu \qquad (1.27)$$

where:
  φ is the solution of homogeneous equation

$$Lu = 0 \qquad (1.28)$$

and the integration constants involved in $f = \varphi + L^{-1}g(t)$ are to be determined by the initial or boundary conditions of the corresponding problem.

The ADM assumes that the unknown function $u(x,t)$ can be expressed by an infinite series of the form

$$u(x,t) = \sum_{n=0}^{\infty} u_n(x,t) \qquad (1.29)$$

and the nonlinear operator Nu can be decomposed by an infinite series of polynomials given by

$$N(u) = \sum_{n=0}^{\infty} A_n \qquad (1.30)$$

where A_n are the Adomian polynomials given by

$$A_n = \frac{1}{n!} \frac{d^n}{d\lambda^n} \left[N \left(\sum_{i=0}^{\infty} \lambda^i u_i \right) \right]_{\lambda=0} \qquad n = 0,1,2,\dots \qquad (1.31)$$

According to the modified decomposition method (MDM) [5], the recursive scheme is given by

$$u_0(x,t) = f_1$$

$$u_1(x,t) = f_2 - L^{-1}Ru_0(x,t) - L^{-1}A_0 \tag{1.32}$$

$$u_{n+1}(x,t) = -L^{-1}Ru_n(x,t) - L^{-1}A_n \quad n \geq 1$$

where the solution of the homogeneous equation (Equation 1.28) is given by

$$f = f_1 + f_2 \tag{1.33}$$

If the zeroth component u_0 is defined, then the remaining components u_n, $n \geq 1$, can be determined completely such that each term is determined in terms of the previous term and the series solution is thus entirely determined. Finally, we approximate the solution $u(x,t)$ by the truncated series.

$$\phi_N(x,t) = \sum_{n=0}^{N-1} u_n(x,t) \quad \text{and} \quad \lim_{N \to \infty} \phi_N(x,t) = u(x,t) \tag{1.34}$$

The method provides the solution in the form of a rapidly convergent series that may lead to the exact solution in the case of linear differential equations and to an efficient numerical solution with high accuracy for nonlinear differential equations. The convergence of the decomposition series has been investigated by several notable researchers [6,7]. The method can significantly reduce the volume of computational work. In comparison with the standard Adomian method, the modified algorithm gives better performance in many cases. The MDM accelerates the convergence of the series solution rapidly if compared with the standard ADM.

1.1.4 Variational Iteration Method

The VIM has been favorably applied to various kinds of nonlinear problems. The main property of the method lies in its flexibility and ability to solve nonlinear equations accurately and conveniently. Major applications to the nonlinear diffusion equation, nonlinear fractional differential equations, nonlinear oscillations, and nonlinear problems arising in various engineering applications are noticed. The VIM was proposed by Ji-Huan He [8,9] and was successfully applied to autonomous ordinary and partial differential equations and other fields. The main advantage of this method is that the correction functions can be constructed easily by the general Lagrange multipliers, which can be optimally determined by the variational theory; the application of restricted variations in the correction functional makes it much easier to determine the multiplier; the initial approximation can be freely selected with possible unknown constants, which can be identified via various methods. The approximations obtained by this method are valid not only for small parameters but also for very large parameters.

To illustrate the basic concept of the VIM [8–11], we consider the following general nonlinear ordinary differential equation:

$$Lu(t) + Nu(t) = g(t) \tag{1.35}$$

where:

L is a linear operator
N is a nonlinear operator
$g(t)$ is a known analytical function

According to He's VIM, we can construct the correction functional as follows:

$$u_{n+1}(t) = u_n(t) + \int\limits_0^t \lambda \left[Lu_n(\xi) + N\tilde{u}_n(\xi) - g(\xi) \right] d\xi \tag{1.36}$$

where:

u_0 is an initial approximation with possible unknowns
λ is a general Lagrange multiplier
$\tilde{u}_n$ is considered as a restricted variation, that is, $\delta\tilde{u}_n = 0$
The Lagrange multiplier λ can be determined from the stationary condition of the correction functional $\delta u_{n+1} = 0$

Being different from other nonlinear analytical methods, such as perturbation methods, the applicability of the VIM does not depend on whether the parameters are small or large; it can find wide application in nonlinear problems without linearization, discretization, or small perturbations.

1.1.5 Explicit Finite Difference Method

Finite difference methods are approximate in the sense that derivatives at a point are approximated by difference quotients over a small interval [12].

Let us consider the following heat conduction equation:

$$\frac{\partial u}{\partial t} = \frac{\partial^2 u}{\partial x^2} \tag{1.37}$$

Using the finite difference approximation method, Equation 1.37 can be discretized to

$$\frac{u_{i,j+1} - u_{i,j}}{k} = \frac{u_{i+1,j} - 2u_{i,j} + u_{i-1,j}}{h^2} \tag{1.38}$$

where:

$$x_i = ih \quad (i = 0, 1, 2, \ldots)$$

and

$$t_j = jk \quad (j = 0, 1, 2, \ldots)$$

Then, Equation 1.38 can be written as

$$u_{i,j+1} = ru_{i-1,j} + (1 - 2r)u_{i,j} + ru_{i+1,j} \tag{1.39}$$

where:

$$r = \frac{\Delta t}{\Delta x^2} = \frac{k}{h^2}$$

Equation 1.39 gives an explicit formula for the unknown *temperature* $u_{i,j+1}$ at the $(i, j+1)$th mesh point in terms of known *temperatures* along the jth time row.
   Similarly, we consider next the fractional diffusion equation [13].

$$\frac{\partial}{\partial t} u(x,t) = K_{\gamma\,0} D_t^{1-\gamma} \frac{\partial^2}{\partial x^2} u(x,t) \tag{1.40}$$

where:
   $_0 D_t^{1-\gamma}$ is the fractional derivative defined through the Riemann–Liouville operator

For any function, $f(t)$ can be expressed in the form of a power series. The fractional derivative of order $1 - \gamma$ at any point inside the convergence region of the power series can be written using the Grünwald–Letnikov fractional derivative in the form of

$$_0 D_t^{1-\gamma} f(t) = \lim_{h \to 0} \frac{1}{h^{1-\gamma}} \sum_{k=0}^{[t/h]} w_k^{(1-\gamma)} f(t - kh) \tag{1.41}$$

where:
   $[t/h]$ refers to the integer part of t/h

The Grünwald–Letnikov definition is simply a generalization of the ordinary discretization formulas for integer order derivatives.
   Using the explicit difference scheme, Equation 1.40 can be discretized to

$$u_j^{m+1} = u_j^m + S_\gamma \sum_{k=0}^{m} w_k^{(1-\gamma)} \left[u_{j-1}^{(m-k)} - 2u_j^{(m-k)} + u_{j+1}^{(m-k)} \right] \tag{1.42}$$

where:
$x_j = j\Delta x$ for $(j = 0,1,2,...)$
$t_m = m\Delta t$ for $(m = 0,1,2,...)$
$u(x_j, t_m) \equiv u_j^{(m)}$ stands for the numerical estimate of the exact value of $u(x,t)$
    at the point (x_j, t_m)

Here, $S_\gamma = K_\gamma \Delta t/[h^{1-\gamma}(\Delta x)^2]$. In this scheme, u_j^{m+1}, for every given position j, is given explicitly in terms of all the previous states u_j^n, $n = 0,1,...,m$.

1.1.6 Differential Transform Method

Many problems in science and engineering can be described by ordinary and partial differential equations. A variety of numerical and analytical methods have been developed to obtain accurate approximate and analytic solutions for the problems in the literature.

The classical Taylor series method is one of the earliest analytic techniques to many problems but it requires a lot of symbolic calculation for the derivatives of functions and higher-order derivatives. The updated version of the Taylor series is called the DTM. The DTM was first proposed by Zhou in 1986 [14]. The DTM is a very effective and powerful solver for various kinds of differential equations. The main advantage of this method is that it can be applied directly to linear and nonlinear differential equations without requiring linearization, discretization, or perturbation.

The differential transform of $f(t)$ can be defined as follows:

$$F(k) = \frac{1}{k!}\left[\frac{d^k f(t)}{dt^k}\right]_{t=t_0} \tag{1.43}$$

Here, $F(k)$ is the transformed function of $f(t)$. The inverse differential transform of $F(k)$ is defined by

$$f(t) = \sum_{k=0}^{\infty} F(k)(t-t_0)^k \tag{1.44}$$

From Equations 1.43 and 1.44, we get

$$f(t) = \sum_{k=0}^{\infty} \frac{(t-t_0)^k}{k!}\left[\frac{d^k f(t)}{dt^k}\right]_{t=t_0} \tag{1.45}$$

This implies that the concept of the differential transform is derived from the Taylor series expansion, but the method does not evaluate the derivatives symbolically.

In this particular case, when $t_0 = 0$, it refers to the Maclaurin series of $f(t)$ and can be expressed as $f(t) = \sum_{k=0}^{\infty} t^k F(k) \equiv DT^{-1}F(k)$ where DT^{-1} stands for inverse differential transform.

The function $f(t)$ can be expressed by a finite series defined as follows:

$$f(t) = \sum_{k=0}^{N} F(k)(t - t_0)^k \tag{1.46}$$

Here, N is the finite sum of terms of the truncated series solution. The DTM is an important tool for solving different classes of nonlinear problems. The fundamental operations of differential transformation are presented in Table 1.1.

If the system considered has a solution in terms of the series expansion of known functions, then using this powerful method, we can obtain the exact solution. The DTM is an effective and reliable tool for the solution of systems of ordinary differential equations. The method gives a rapidly convergent series solution. The accuracy of the obtained solution can be improved by taking more terms in the solution. In many cases, the series solution obtained with the DTM can be written in the exact closed form. This method reduces the computational difficulties involved in the other traditional methods, and all the calculations can be done easily and efficiently.

TABLE 1.1

The Fundamental Operations of Differential Transformation

Properties	Time Function	Transformed Function	
1	$f(t) = \alpha u(t) + \beta v(t)$	$F(k) = \alpha U(k) + \beta V(k)$	
2	$f(t) = u(t)v(t)$	$F(k) = \sum_{l=0}^{k} U(l)V(k-l)$	
3	$f(t) = \dfrac{du(t)}{dt}$	$F(k) = (k+1)U(k+1)$	
4	$f(t) = \dfrac{d^m u(t)}{dt^m}$	$F(k) = \dfrac{(k+m)!}{k!}U(k+m)$	
5	$f(t) = t^m$	$F(k) = \dfrac{1}{k!}\dfrac{d^k\left[f(t)\right]}{dt^k}\bigg|_{t=0} = \delta(k-m) = \begin{cases} 1, & k = m \\ 0, & k \neq m \end{cases}$	
		$F(k) = \dfrac{1}{k!}\dfrac{d^k\left[f(t)\right]}{dt^k}\bigg|_{t=t_i} = \dfrac{m(m-1)\cdots(m-k+1)t_i^{m-k}}{k!}$	
6	$f(t) = \sin(\omega t + \alpha)$	$F(k) = \dfrac{1}{k!}\dfrac{d^k\left[f(t)\right]}{dt^k}\bigg|_{t=0} = \dfrac{\omega^k}{k!}\sin\left(\dfrac{k\pi}{2} + \alpha\right)$	
		$F(k) = \dfrac{1}{k!}\dfrac{d^k\left[f(t)\right]}{dt^k}\bigg|_{t=t_i} = \dfrac{\omega^k}{k!}\sin\left(\dfrac{k\pi}{2} + \omega t_i + \alpha\right)$	

1.1.7 Multistep DTM

The DTM is always used to provide approximate solutions for a class of non-linear problems in terms of a convergent series with easily computable components. But it has a drawback: the series solution converges into a very small region and it has slow convergent rate in the wider regions. To overcome this problem, we represent in this section the MDTM. Let us consider the following nonlinear initial value problem:

$$f\left(t, u, u', \ldots, u^{(p)}\right) = 0 \quad u^{(p)} \text{ is the } p\text{th derivative of } u \qquad (1.47)$$

subject to the initial conditions $u^{(k)}(0) = c_k$, for $k = 0,1,\ldots,p-1$.

We find the solution over the interval $[0,T]$. The approximate solution of the initial value problem can be expressed by the finite series

$$u(t) = \sum_{m=0}^{M} a_m t^m \quad t \in [0, T] \qquad (1.48)$$

Assume that the interval $[0,T]$ is divided into N subintervals $[t_{n-1}, t_n]$, with $n = 1,2,\ldots,N$ of equal step size $h = T/N$ using the node point $t_n = nh$. The main idea of the multistep DTM is to apply first the DTM to Equation 1.47 over the interval $[0, t_1]$; we obtain the following approximate solutions [15]:

$$u_1(t) = \sum_{m=0}^{M_1} a_{1m} t^m, \quad t \in [0, t_1] \qquad (1.49)$$

using the initial conditions $u_1^{(k)}(0) = c_k$. For $n \geq 2$ and at each subinterval $[t_{n-1}, t_n]$, we use the initial conditions $u_n^{(k)}(t_{n-1}) = u_{n-1}^{(k)}(t_{n-1})$ and apply the DTM to Equation 1.47 over the interval $[t_{n-1}, t_n]$, where t_0 in Equation 1.43 is replaced by t_{n-1}. This process is repeated and generates a series of approximate solutions $u_n(t)$, $n = 1,2,\ldots,N$. Now

$$u_n(t) = \sum_{m=0}^{M_1} a_{nm} \left(t - t_{n-1}\right)^m, \quad t \in [t_n, t_{n+1}] \qquad (1.50)$$

where $M = M_1 \cdot N$. Hence, the multistep DTM assumes the following solution:

$$u(t) = \begin{cases} u_1(t), & t \in [0, t_1] \\ u_2(t), & t \in [t_1, t_2] \\ \vdots \\ u_N(t), & t \in [t_{N-1}, t_N] \end{cases} \qquad (1.51)$$

The multistep DTM [15–18] is used for computational techniques for all values of h. It can be easily shown that if step size $h = T$, then the multistep DTM reduces to the classical DTM. The main advantage of this new algorithm is that the obtained series solution converges for wide time regions.

These analytical methods and numerical techniques are the most vital solvers for ordinary as well as partial differential equations, linear and non-linear problems, and fractional differential equations appearing in the field of science and engineering. The numerical methods are an iterative process, by using which we can obtain the solution more accurately; accuracy can be further improved when the step size of each subinterval becomes smaller. The approximate solutions obtained by the analytical methods are presented by a truncated form of infinite series.

1.1.8 Fractional DTM

The DTM was first applied in the engineering domain by Zhou [14]. This method is a numerical method based on the Taylor series expansion, which constructs an analytical solution in the form of a polynomial. The tradi-tional higher-order Taylor series method requires symbolic computation. The Taylor series method is computationally taken long time for large order derivatives. The DTM is an iterative procedure for obtaining the analytic Taylor series solution of ordinary or partial differential equations. In this sec-tion, we define the fractional DTM (FDTM), which is based on the general-ized Taylor series formula:

$$F_\alpha(k) = \frac{1}{\Gamma(\alpha k + 1)} \left[\left(D_{x_0}^\alpha \right)^k f(x) \right]_{x=x_0} \tag{1.52}$$

where $\left(D_{x_0}^\alpha \right)^k \equiv D_{x_0}^\alpha . D_{x_0}^\alpha ... D_{x_0}^\alpha$ k-times, and the differential inverse transform of $F_\alpha(k)$ is defined as follows:

$$f(x) = \sum_{k=0}^{\infty} F_\alpha(k)(x - x_0)^{\alpha k} \tag{1.53}$$

If we substitute Equation 1.52 into Equation 1.53, we obtain

$$\sum_{k=0}^{\infty} F_\alpha(k)(x - x_0)^{\alpha k} = \sum_{k=0}^{\infty} \frac{(x - x_0)^{\alpha k}}{\Gamma(\alpha k + 1)} \left\{ \left[\left(D_{x_0}^\alpha \right)^k f \right](x_0) \right\} = f(x)$$

Hence, Equation 1.53 is the inverse fractional differential transform of Equation 1.52.

For the case $\alpha = 1$, then the fractional differential transform in Equation 1.52 reduces to the classical differential transform.

1.2 Numerical Methods for Solving Stochastic Point Kinetic Equations

1.2.1 Wiener Process or Brownian Motion Process

A standard Wiener process (often called Brownian motion) on the interval $[0,T]$ is a continuous time stochastic process $W(t)$ that depends continuously on $t \in [0,T]$ and satisfies the following properties [19–21]:

1. $W(0) = 0$ (with probability 1).
2. For $0 \le s \le t \le T$, the increment $W(t) - W(s)$ is normally distributed with mean $E[W(t)] = 0$, variance $E[W(t) - W(s)]^2 = |t - s|$, and covariance $E[W(t)W(s)] = \min(t,s)$; equivalently $W(t) - W(s) \sim \sqrt{t - s}N(0,1)$, where $N(0,1)$ denotes a normal distribution with zero mean and unit variance.
3. For $0 \le s < t < u < v \le T$, the increments $W(t) - W(s)$ and $W(v) - W(u)$ are independent. For computational purposes, it is useful to consider discredited Brownian motion, where $W(t)$ is specified at discrete t values. We thus set $\Delta t = T/N$ for some positive integer N and let $W_i = W(t_i)$ with $t_i = i\Delta t$. We discretize the Wiener process with time-step Δt as $W_i = W_{i-1} + dW_i$, $i = 1,2,...,N$, where each $dW_i \sim \sqrt{\Delta t}N(0,1)$.

Stochastic differential equation (SDE) models play a prominent role in many areas, including biology, chemistry, epidemiology, mechanics, microelectronics, economics, and finance.

An Itô process (or stochastic integral) $X = \{X_t, t \ge 0\}$ has the following form [19–21]:

$$X_t = X_0 + \int_0^t a(X_s)ds + \int_0^t b(X_s)dW_s, \text{ for } t \ge 0 \tag{1.54}$$

It consists of an initial value $X_0 = x_0$, which may be random, a slowly varying continuous component called the drift, and a rapidly varying continuous random component called the diffusion. The second integral in Equation 1.54 is an Itô stochastic integral with respect to the Wiener process $W = \{W_t, t \ge 0\}$. The integral equation in Equation 1.54 is often written in the following differential form:

$$dX_t = a(X_t)dt + b(X_t)dW_t \tag{1.55}$$

Equation 1.55 represents the SDE (or Itô SDE). Here, we describe the Euler–Maruyama method and the strong order 1.5 Taylor methods to simulate a stochastic point kinetic equation.

1.2.2 Euler–Maruyama Method

The Euler–Maruyama approximation is the simplest time discrete approximation of an Itô process. Let $\{Y_\tau\}$ be an Itô process on $\tau \in [t_0, T]$ satisfying the SDE.

$$\begin{cases} dY_\tau = a(\tau, Y_\tau) d\tau + b(\tau, Y_\tau) dW_\tau \\ Y_{t_0} = Y_0 \end{cases} \tag{1.56}$$

For a given time discretization

$$t_0 = \tau_0 < \tau_1 < \cdots < \tau_n = T \tag{1.57}$$

an Euler approximation is a continuous time stochastic process $\{X(\tau), t_0 \leq \tau \leq T\}$ satisfying the following iterative scheme [20,21]:

$$X_{n+1} = X_n + a(\tau_n, X_n) \Delta \tau_{n+1} + b(\tau_n, X_n) \Delta W_{n+1} \tag{1.58}$$

for $n = 0,1,2,\ldots,N-1$, with the initial value

$$X_0 = X(\tau_0)$$

where:

$$X_n = X(\tau_n), \ \Delta \tau_{n+1} = \tau_{n+1} - \tau_n$$
$$\Delta W_{n+1} = W(\tau_{n+1}) - W(\tau_n)$$

Here, each random number ΔW_n is computed as $\Delta W_n = \eta_n \sqrt{\Delta \tau_n}$, where η_n is chosen from the standard normal distribution $N(0,1)$.

We have considered the equidistant discretized times $\tau_n = \tau_0 + n\Delta$ with $\Delta = \Delta_n = (T - \tau_0)/N$ for some integer N large enough so that $\Delta \in (0,1)$.

1.2.3 Strong Order 1.5 Taylor Method

Here we consider the Taylor approximation having strong order $\alpha = 1.5$. The strong order 1.5 Taylor scheme can be obtained by adding more terms from the Itô–Taylor expansion to the Milstein scheme [20,21]. The strong order 1.5 Itô–Taylor scheme is

$$Y_{n+1} = Y_n + a\Delta_n + b\Delta W_n + \frac{1}{2} bb_x \left(\Delta W_n^2 - \Delta_n \right) + a_x b \Delta Z_n + \frac{1}{2} \left(aa_x + \frac{1}{2} b^2 a_{xx} \right) \Delta_n^2$$

$$+ \left(ab_x + \frac{1}{2} b^2 b_{xx} \right) \left(\Delta W_n \Delta_n - \Delta Z_n \right) + \frac{1}{2} b \left(bb_{xx} + b_x^2 \right) \left(\frac{1}{3} \Delta W_n^2 - \Delta_n \right) \Delta W_n \tag{1.59}$$

for $n = 0,1,2,\ldots,N-1$, with the initial value

$$Y_0 = Y(\tau_0) \text{ and } \Delta_n = \Delta\tau_n$$

Here, partial derivatives are denoted by subscripts and the random variable ΔZ_n is normally distributed with mean $E(\Delta Z_n)=0$ and variance $E(\Delta Z_n^2)=(1/3)\Delta\tau_n^3$ and correlated with ΔW_n by covariance

$$E(\Delta Z_n \Delta W_n) = \frac{1}{2}\Delta\tau_n^2$$

We can generate ΔZ_n as

$$\Delta Z_n = \frac{1}{2}\Delta\tau_n\left(\Delta W_n + \frac{\Delta V_n}{\sqrt{3}}\right) \qquad (1.60)$$

where ΔV_n is chosen independently from $\sqrt{\Delta\tau_n}\ N(0,1)$. Here, the approximation $Y_n = Y(\tau_n)$ is the continuous time stochastic process $\{Y(\tau), t_0 \le \tau \le T\}$. The time step-size $\Delta\tau_n = \tau_n - \tau_{n-1}$ and $\Delta W_n = W(\tau_n) - W(\tau_{n-1})$.

2

Neutron Diffusion Equation Model in Dynamical Systems

2.1 Introduction

The nuclear reactor forms the heart of a nuclear power plant. Fundamental to a nuclear reactor are nuclear physics and reactor physics, which deal with the basic aspects of the design of nuclear reactors. This knowledge is essential for understanding reactor behavior during normal operation as well as abnormal conditions. Nuclear engineering is an excellent technology by which tremendous amounts of energy is generated from a small amount of fuel. In addition to power generation, numerous applications are expected in the future. In addition to being used in energy generation, neutrons are expected to be widely used as a medium in nuclear reactions. Here, nuclear energy refers to the energy released in nuclear fission. This occurs because of the absorption of neutrons by fissile material. Neutrons are released by nuclear fission, and since the number of neutrons released is sufficiently greater than 1, a chain reaction of nuclear fission can be established. This allows, in turn, for energy to be extracted from the process. The amount of extracted energy can be adjusted by controlling the number of neutrons. The higher the power density, the greater the economic efficiency of the reactor. Ultimately, this means careful control of neutron distribution. If there is an accident in a reactor system, the power output will run out of control. This situation is almost the same as an increase in the number of neutrons. Thus, the theory of nuclear reactors can be considered to be the study of the behavior of neutrons in a nuclear reactor. The design of nuclear reactors is such that there is a balance between the production of neutrons in fission reactions and the loss of neutrons due to capture or leakage. The study of such a process is known as nuclear reactor theory or nuclear reactor physics [22–24].

In nuclear physics, for the purpose of optimizing the performance and regulating the safety of a nuclear reactor, it is important that the nuclear reactor run at a critical level. To describe the state of criticality, we must understand the nature of nuclear power. Nuclear power is based upon a process called fission, a process in which a neutron approaches a fissile isotope, and its very proximity, as the neutron slows near the atom, causes it to split into two or more pieces, generating fission products and even more neutrons, called prompt and delayed neutrons. These neutrons collide with hydrogen in the water surrounding the fuel pins, depositing their energy and increasing the temperature of the water, causing it to boil. The heat of the water or rather the stream is then used to power turbines and generate power.

Therefore, neutron diffusion equations (NDE) as well as neutron point kinetic equations have been analyzed to study the population of neutron density in the system and precursor density, that is, the population of fission products that results in delayed neutrons.

2.2 Outline of the Present Study

In this chapter, we apply the effective analytical and numerical methods discussed in Chapter 1 to obtain the solution for the NDE. The neutrons are here characterized by a single energy or speed, and the model allows preliminary design estimates.

Now, we consider the time-independent fixed source one-group NDE [22] for a homogeneous region where the geometry with the vacuum boundary conditions are valid and is given by

$$\nabla^2 \phi(\vec{r}) - \kappa^2 \phi(\vec{r}) = -\frac{S(\vec{r})}{D}, \vec{r} \in V, \phi(\vec{r}) = 0, \vec{r} \in S \tag{2.1}$$

where:
$\phi(\vec{r})$ is the neutron flux
$S(\vec{r})$ is the neutron source
Σ_a is the absorption cross section
D is the diffusion constant, given by inverse diffusion length $\kappa^2 = \Sigma_a/D$

We consider the NDE with the fixed source for a two-dimensional system with a square geometry; it is symmetric with respect to both x and y axes. In this scenario, the NDE together with the boundary conditions given by Equation 2.1 reduces to

$$\frac{\partial^2 \phi(x,y)}{\partial x^2} + \frac{\partial^2 \phi(x,y)}{\partial y^2} - \kappa^2 \phi(x,y) = -\frac{S}{D} \tag{2.2}$$

$$\left.\frac{\partial \phi(x,y)}{\partial x}\right|_{x=0} = 0, \quad \phi(x,y)\big|_{x=a} = 0 \tag{2.3}$$

$$\left.\frac{\partial\phi(x,y)}{\partial y}\right|_{y=0} = 0, \quad \phi(x,y)\big|_{y=a} = 0 \tag{2.4}$$

2.3 Application of the Variational Iteration Method to Obtain the Analytical Solution of the NDE

The variational iteration method (VIM) is a powerful method to investigate approximate solutions. It is based on the incorporation of a general Lagrange multiplier in the construction of correction functional for the equation. In addition, no linearization and perturbation is required by the method. The VIM method, which is also known as the modified general Lagrange's multiplier method [8,25], has been shown to solve effectively, easily, and accurately a large class of nonlinear models of real physical problems.

In this section, in order to solve the NDE (Equation 2.1), we apply the VIM. To illustrate the basic concept of the VIM [8,9] we consider the following general nonlinear ordinary differential equation given by

$$Lu(t) + Nu(t) = g(t)$$

where:
L is a linear operator
N is a nonlinear operator
$g(t)$ is a known analytical function

According to He's VIM, we can construct the correction functional as follows:

$$u_{n+1}(t) = u_n(t) + \int_0^t \lambda\left[Lu_n(\xi) + N\tilde{u}_n(\xi) - g(\xi)\right]d\xi$$

where:
u_0 is an initial approximation with possible unknowns
λ is a general Lagrange multiplier
$\tilde{u}_n$ is considered as a restricted variation, that is, $\delta\tilde{u}_n = 0$

The Lagrange multiplier λ can be determined from the stationary condition of the correction functional $\delta u_{n+1} = 0$.

Now, to solve the NDE (Equation 2.1) we construct a correction functional as follows:

$$\phi_{n+1}(x,y) = \phi_n(x,y) + \int_0^x \lambda(x')\left[\frac{\partial^2\phi_n(x',y)}{\partial x'^2} + \frac{\partial^2\tilde{\phi}_n(x',y)}{\partial y^2} - \kappa^2\tilde{\phi}_n(x',y) + \frac{S}{D}\right]dx' \tag{2.5}$$

where:

λ is a general Lagrange multiplier [26], which can be identified optimally via variational theory

$\phi(x,0)$ is an initial approximation with possible unknowns

$\tilde{\phi}_n(x',y)$ is considered as the restricted variation, that is, $\delta\tilde{\phi}_n = 0$

Making the above correction functional stationary and to find the optimal value of λ, we have

$$\delta\phi_{n+1}(x,y) = \delta\phi_n(x,y)$$

$$+ \delta\int_0^x \lambda(x')\left[\frac{\partial^2\phi_n(x',y)}{\partial x'^2} + \frac{\partial^2\tilde{\phi}_n(x',y)}{\partial y^2} - \kappa^2\tilde{\phi}_n(x',y) + \frac{S}{D}\right]dx' = 0 \qquad (2.6)$$

which yields the following stationary conditions

$$\ddot{\lambda}(x') = 0 \qquad (2.7)$$

$$\lambda(x')\big|_{x'=x} = 0 \qquad (2.8)$$

$$1 - \dot{\lambda}(x')\big|_{x'=x} = 0 \qquad (2.9)$$

Equation 2.7 is called the Lagrange–Euler equation and Equations 2.8 and 2.9 are the natural boundary conditions.

The Lagrange multiplier can, therefore, be identified as

$$\lambda(x') = x' - x \qquad (2.10)$$

Considering the boundary conditions in Equation 2.3, we assume an initial approximation in the form of infinite cosine series:

$$\phi_0(x,y) = \sum_{i=0}^\infty b_i \cos(\beta_i y) \qquad (2.11)$$

Now, Equation 2.11 satisfying the boundary condition at $y = a$ yields

$$\beta_n = \frac{(2n+1)\pi}{2a}, \qquad n = 0,1,2,\dots \qquad (2.12)$$

For convenience, we consider the known force term in the same basis set with Equation 2.11, that is,

$$\frac{S}{D} = \sum_{i=0}^{\infty} b_i \cos(\beta_i y) \tag{2.13}$$

where the orthogonalty property of the Fourier basis yields

$$b_i = \frac{(-1)^i 2S}{aD\beta_i} \tag{2.14}$$

Using Equation 2.10 and the initial approximation Equation 2.11, the first approximation becomes

$$\phi_1(x,y) = \phi_0(x,y) + \int_0^x (x'-x) \left[\frac{\partial^2 \phi_0(x',y)}{\partial x'^2} + \frac{\partial^2 \phi_0(x',y)}{\partial y^2} - \kappa^2 \phi_0(x',y) + \frac{S}{D} \right] dx'$$

$$= \sum_{i=0}^{\infty} b_i \cos(\beta_i y)$$

$$+ \int_0^x (x'-x) \left[\frac{\partial^2 \phi_0(x',y)}{\partial x'^2} + \frac{\partial^2 \phi_0(x',y)}{\partial y^2} - \kappa^2 \phi_0(x',y) + \sum_{i=0}^{\infty} b_i \cos(\beta_i y) \right] dx' \tag{2.15}$$

$$= \sum_{i=0}^{\infty} b_i \cos(\beta_i y) \left[1 + (\alpha_i^2 - 1)\frac{x^2}{2!} \right], \quad \text{where } \alpha_i^2 = \beta_i^2 + \kappa^2$$

The second approximation is

$$\phi_2(x,y) = \phi_1(x,y) + \int_0^x (x'-x) \left[\frac{\partial^2 \phi_1(x',y)}{\partial x'^2} + \frac{\partial^2 \phi_1(x',y)}{\partial y^2} - \kappa^2 \phi_1(x',y) + \frac{S}{D} \right] dx'$$

$$= \sum_{i=0}^{\infty} b_i \cos(\beta_i y) \left[1 + (\alpha_i^2 - 1)\frac{x^2}{2!} \right]$$

$$+ \int_0^x (x'-x) \left[\frac{\partial^2 \phi_1(x',y)}{\partial x'^2} + \frac{\partial^2 \phi_1(x',y)}{\partial y^2} - \kappa^2 \phi_1(x',y) + \sum_{i=0}^{\infty} b_i \cos(\beta_i y) \right] dx' \tag{2.16}$$

$$= \sum_{i=0}^{\infty} b_i \cos(\beta_i y) \left[1 + (\alpha_i^2 - 1)\frac{x^2}{2!} + \frac{\alpha_i^4 x^4}{4!} - \frac{\alpha_i^2 x^4}{4!} \right]$$

The third approximation is

$$
\phi_3(x,y)= \phi_2(x,y)+ \int_0^x (x'-x)\left[\frac{\partial^2\phi_2(x',y)}{\partial x'^2}+\frac{\partial^2\phi_2(x',y)}{\partial y^2}-\kappa^2\phi_2(x',y)+\frac{S}{D}\right]dx'
$$

$$
= \sum_{i=0}^{\infty} b_i \cos(\beta_i y)\left[1+\left(\alpha_i^2-1\right)\frac{x^2}{2!}+\frac{\alpha_i^4 x^4}{4!}-\frac{\alpha_i^2 x^4}{4!}\right]
$$

(2.17)

$$
+\int_0^x (x'-x)\left[\frac{\partial^2\phi_2(x',y)}{\partial x'^2}+\frac{\partial^2\phi_2(x',y)}{\partial y^2}-\kappa^2\phi_2(x',y)+\sum_{i=0}^{\infty} b_i \cos(\beta_i y)\right]dx'
$$

$$
= \sum_{i=0}^{\infty} b_i \cos(\beta_i y)\left[1+\left(\alpha_i^2-1\right)\frac{x^2}{2!}+\frac{\alpha_i^4 x^4}{4!}-\frac{\alpha_i^2 x^4}{4!}+\frac{\alpha_i^6 x^6}{6!}-\frac{\alpha_i^4 x^6}{6!}\right]
$$

Similarly, the rest of the approximations of the iteration results can be obtained. Here the VIM has been successfully applied to find the approximate solution to the one-group NDE.

2.4 Application of the Modified Decomposition Method to Obtain the Analytical Solution of the NDE

In this section, we use the modified decomposition method (MDM) to obtain the approximate solution of one-group NDE (Equation 2.1). Large classes of linear and nonlinear differential equations, ordinary as well as partial, can be solved by the MDM. A reliable modification of the Adomian decomposition method (ADM) has been done by Wazwaz [5]. The decomposition method provides an effective procedure for the analytical solution of a wide and general class of dynamical systems representing real physical problems [27–29]. This method efficiently works for initial-value or boundary-value problems and for linear or nonlinear, ordinary, or partial differential equations. Moreover, we have the advantage of a single global method, viz. the MDM, for solving ordinary or partial differential equations as well as many types of other equations.

We rewrite Equation 2.2 in the following operator form:

$$
L_x\phi(x,y)=-\frac{S}{D}-L_y\{\phi(x,y)\}+\kappa^2\phi(x,y)
$$

(2.18)

where:

$$
L_x \equiv \frac{\partial^2}{\partial x^2}
$$

$$L_y \equiv \frac{\partial^2}{\partial y^2}$$

Now we apply the twofold integration inverse operator L_x^{-1} to Equation 2.18, and we obtain

$$L_x^{-1}L_x\phi(x,y) = -L_x^{-1}\left[\frac{S}{D}\right] - L_x^{-1}\left[L_y\{\phi(x,y)\} + L_x^{-1}\{\kappa^2\phi(x,y)\}\right] \qquad (2.19)$$

$$\phi(x,y) = f(x,y) - L_x^{-1}\left[\frac{S}{D}\right] - L_x^{-1}\left[L_y\{\phi(x,y)\} + L_x^{-1}\{\kappa^2\phi(x,y)\}\right] \qquad (2.20)$$

where $f(x,y)$ is the solution of $L_x\phi(x,y) = 0$. From MDM methodology [5], we assume the infinite series solution for $\phi(x,y)$ as

$$\phi(x,y) = \sum_{n=0}^{\infty} \phi_n(x,y)$$

From the recursive scheme equation (Equation 1.33) of MDM, we obtain

$$\phi_0(x,y) = f \qquad (2.21)$$

$$\phi_1(x,y) = -L_x^{-1}\left[\frac{S}{D}\right] - L_x^{-1}\left[L_y\{\phi_0(x,y)\} + L_x^{-1}\{\kappa^2\phi_0(x,y)\}\right] \qquad (2.22)$$

$$\phi_2(x,y) = -L_x^{-1}\left[L_y\{\phi_1(x,y)\} + L_x^{-1}\{\kappa^2\phi_1(x,y)\}\right] \qquad (2.23)$$

In general $\phi_{n+1}(x,y) = -L_x^{-1}[L_y\{\phi_n(x,y)\}] + L_x^{-1}[\kappa^2\phi_n(x,y)]$, for all $n \geq 1$.

Considering the boundary conditions, we assume an initial approximation in the form of Fourier cosine series:

$$\phi_0(x,y) = \sum_{i=0}^{\infty} b_i \cos(\beta_i y) \qquad (2.24)$$

Now Equation 2.24 satisfying the boundary conditions at $y = a$ yields

$$\beta_n = \frac{(2n+1)\pi}{2a}, \quad n = 0,1,2,\ldots \qquad (2.25)$$

For convenience, we consider the known force term in the same basis set with Equation 2.24, that is,

$$\frac{S}{D} = \sum_{i=0}^{\infty} b_i \cos(\beta_i y) \qquad (2.26)$$

where the orthogonalty property of Fourier basis yields

$$s_i = \frac{(-1)^i 2S}{aD\beta_i} \tag{2.27}$$

$$\phi_1(x,y) = -L_x^{-1}\left[\frac{S}{D}\right] - L_x^{-1}\left[L_y\left\{\phi_0(x,y)\right\}\right] + L_x^{-1}\left[\kappa^2\phi_0(x,y)\right]$$

$$= -L_x^{-1}\left[\sum_{i=0}^{\infty} s_i \cos(\beta_i y)\right] - L_x^{-1}L_y\left[\sum_{i=0}^{\infty} b_i \cos(\beta_i y)\right]$$

$$+ L_x^{-1}\left[\kappa^2\sum_{i=0}^{\infty} b_i \cos(\beta_i y)\right]$$

$$= \sum_{i=0}^{\infty} b_i \cos(\beta_i y)\left(\frac{\alpha_i^2 x^2}{2!}\right) - \sum_{i=0}^{\infty} s_i \cos(\beta_i y)\frac{x^2}{2!}, \quad \text{where } \alpha_i^2 = \beta_i^2 + \kappa^2 \tag{2.28}$$

$$\phi_2(x,y) = -L_x^{-1}\left[L_y\left\{\phi_1(x,y)\right\}\right] + L_x^{-1}\left[\kappa^2\phi_1(x,y)\right]$$

$$= -L_x^{-1}L_y\left[\sum_{i=0}^{\infty} b_i \cos(\beta_i y)\left(\frac{\alpha_i^2 x^2}{2!}\right) - \sum_{i=0}^{\infty} s_i \cos(\beta_i y)\frac{x^2}{2!}\right]$$

$$+ \kappa^2 L_x^{-1}\left[\sum_{i=0}^{\infty} b_i \cos(\beta_i y)\left(\frac{\alpha_i^2 x^2}{2!}\right) - \sum_{i=0}^{\infty} s_i \cos(\beta_i y)\frac{x^2}{2!}\right]$$

$$= \sum_{i=0}^{\infty} b_i \cos(\beta_i y)\left(\frac{\alpha_i^4 x^4}{4!}\right) - \sum_{i=0}^{\infty} s_i \cos(\beta_i y)\frac{\alpha_i^2 x^4}{4!}$$

and so on.

In this manner, we can completely determine all the remaining components of the infinite series. Therefore, the solution of the NDE is given by

$$\phi(x,y) = \sum_{i=0}^{\infty} \phi_i(x,y)$$

$$= \sum_{i=0}^{\infty} b_i \cos(\beta_i y)\cosh(\alpha_i x) + \sum_{i=0}^{\infty} \frac{s_i}{\alpha_i^2}\cos(\beta_i y)\left[1 - \cosh(\alpha_i x)\right] \tag{2.29}$$

Now, applying the boundary condition at $x = a$, we obtain

$$b_i \cosh(\alpha_i a) + \frac{s_i}{\alpha_i^2}\left[1 - \cosh(\alpha_i a)\right] = 0, \quad i = 0,1,2,\cdots \tag{2.30}$$

This implies that

$$b_i = \frac{(-1)^i 2S}{aD\beta_i \alpha_i^2}\left[\frac{\cosh(\alpha_i a)-1}{\cosh(\alpha_i a)}\right], \quad i = 0,1,2,\cdots \tag{2.31}$$

Consequently, we obtain

$$1-\alpha_i^2 = \frac{1}{\cosh(\alpha_i a)}, \quad i = 0,1,2,\cdots \tag{2.32}$$

Substituting the value of s_i given by Equation 2.27 together with Equation 2.32 in Equation 2.29 yields

$$\phi(x,y) = \frac{2S}{aD}\sum_{i=0}^{\infty}\frac{(-1)^i}{\beta_i\alpha_i^2}\cos(\beta_i y)\left[1-\frac{\cosh(\alpha_i x)}{\cosh(\alpha_i a)}\right] \tag{2.33}$$

which is an exact solution [30]. In practical computation, we take a three-term approximation to $\phi(x,y)$, namely,

$$\Phi(x,y) = \phi_0 + \phi_1 + \phi_2$$

$$= \sum_{i=0}^{\infty}b_i\cos(\beta_i y)\left[1+(\alpha_i^2-1)\frac{x^2}{2!}+\frac{\alpha_i^4 x^4}{4!}-\frac{\alpha_i^2 x^4}{4!}\right] \tag{2.34}$$

The MDM accelerates the convergence of the series solution rapidly, dramatically reduces the volume of work, and provides an exact solution by using fewer iterations. In view of the present analysis, the MDM can be employed for the solution of the NDE with fixed source [31].

2.5 Numerical Results and Discussions for NDEs

In this analysis, we consider a square reactor core with edge length $2a = 50$ cm and apply the VIM and MDM to obtain a numerical solution for one quadrant of the system. This is sufficient owing to the symmetricity regarding the vacuum conditions at the left and upper boundaries together with the reflector conditions at the right and lower boundaries, as expressed in Equation 2.2.

In this numerical discussion, we assume the constant parameters of the reactor as presented in Table 2.1.

We present the numerically computed result obtained by the VIM and MDM in Figure 2.1 for $y = 0$ and the neutron flux distribution $\phi(x,y)$ in Figure 2.2,

TABLE 2.1

Constants of the Reactor

Constants	a (cm)	D (cm)	Σ_a (cm^{-1})	S
Value	25	1.77764	1	1

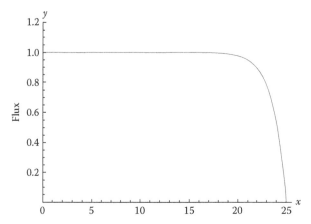

FIGURE 2.1
Neutron flux for $y = 0$.

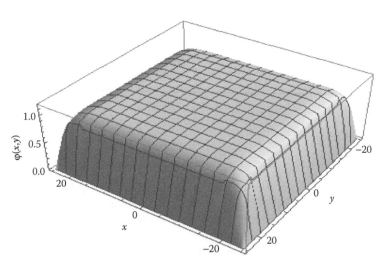

FIGURE 2.2
Neutron flux distribution.

respectively. The obtained results indicate that the two analytical methods like VIM and MDM, compared to the widely used analytic method of separation of variables, yields efficient and relatively straightforward expressions for the solution of the NDE.

2.6 One-Group NDE in Cylindrical and Hemispherical Reactors

In this section, the nonlinear analytical homotopy analysis method (HAM) and ADM have been implemented to solve the one-group NDE [22,32] in hemispherical and cylindrical reactors. The present analysis provides the application of the HAM and ADM to compute the critical radius and flux distribution of the time-independent NDE for both symmetrical bodies. The different boundary conditions are utilized like zero flux at boundary as well as the zero flux at extrapolated boundary. The process of flux distribution takes place in two symmetrical reactors (see Figure 2.3). Figure 2.3a represents a finite cylinder having height h and radius a. Figure 2.3b represents an hemispherical geometry of radius a, where the flux in the reactor is a function of both r and θ.

2.6.1 Application of HAM to a Cylindrical Reactor

The HAM is used to formulate a new analytic solution of the NDE for cylindrical reactors. The method [1,33,34] has proved useful for problems involving algebraic, linear/nonlinear, and ordinary/partial differential equations. Being an analytic recursive method, it provides a series sum solution. It has

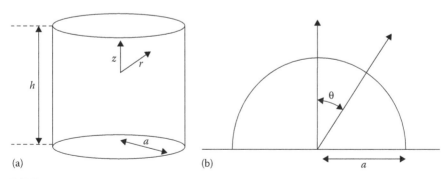

(a)   (b)

FIGURE 2.3
Flux distribution of symmetrical reactors through (a) cylindrical reactor and (b) hemisphere geometry.

the advantage of offering us freedom to choose its arguments, such as the initial guess, the auxiliary linear operator, and the convergence control parameter. Further it allows us to effectively control the rate and region of convergence of the series solution.

To illustrate the HAM, we consider the following differential equation:

$$N[u(x,t)] = 0 \qquad (2.35)$$

where:
 N is the nonlinear operator
 x and t are the independent variables
 $u(x,t)$ is an unknown function

For simplicity, we ignore all boundary or initial conditions, which are treated in the same way. By means of generalizing the HAM [1,33,34], we first construct the zeroth-order deformation equation:

$$(1-q)L\big[\phi(x,t;q) - u_0(x,t)\big] = qhH(x,t)N\big[\phi(x,t;q)\big] \qquad (2.36)$$

where:
 $q \in [0,1]$ is the embedding parameter
 $h \neq 0$ is an auxiliary parameter
 L is an auxiliary linear operator
 $\phi(x,t;q)$ is an unknown function
 $u_0(x,t)$ is an initial guess of $u(x,t)$
 $H(x,t)$ is a nonzero auxiliary function

For $q = 0$ and $q = 1$, the zeroth-order deformation equation given by Equation 2.36 leads to

$$\phi(x,t;0) = u_0(x,t) \text{ and } \phi(x,t;1) = u(x,t) \qquad (2.37)$$

When the value of q increases from 0 to 1, the solution $\phi(x,t;q)$ varies from the initial guess $u_0(x,t)$ to the solution $u(x,t)$. Expanding $\phi(x,t;q)$ in Taylor's series with respect to q we have

$$\phi(x,t;q) = u_0(x,t) + \sum_{m=1}^{\infty} u_m(x,t)q^m \qquad (2.38)$$

where

$$u_m(x,t) = \frac{1}{m!} \frac{\partial^m \phi(x,t;q)}{\partial q^m}\bigg|_{q=0} \qquad (2.39)$$

The convergence of the series (Equation 2.38) depends upon the auxiliary parameter h. If it is convergent at $q = 1$, we get

$$u(x,t) = u_0(x,t) + \sum_{m=1}^{\infty} u_m(x,t) \tag{2.40}$$

which must be one of the solutions of the original differential equation. Now we define a vector

$$\vec{u}_m(x,t) = \{u_0(x,t), u_1(x,t), u_2(x,t), \cdots, u_m(x,t)\} \tag{2.41}$$

Differentiating the zeroth-order deformation equation (Equation 2.36) m times with respect to q, then dividing them by $m!$, and finally setting $q = 0$, we obtain the following mth-order deformation equation

$$L[u_m(x,t) - \chi_m u_{m-1}(x,t)] = hH(x,t)\mathfrak{R}_m[\vec{u}_{m-1}(x,t)] \tag{2.42}$$

where

$$\mathfrak{R}_m(\vec{u}_{m-1}) = \frac{1}{(m-1)!} \frac{\partial^{m-1} N[\phi(x,t;q)]}{\partial q^{m-1}}\bigg|_{q=0} \tag{2.43}$$

and

$$\chi_m = \begin{cases} 0, & m \leq 1 \\ 1, & m > 1 \end{cases} \tag{2.44}$$

Now, the solution for the mth-order deformation equation (Equation 2.42), after applying L^{-1} on both sides, is

$$u_m(x,t) = \chi_m u_{m-1}(x,t) + L^{-1}\{hH(x,t)\mathfrak{R}_m[u_{m-1}(x,t)]\} \tag{2.45}$$

In this way, it is easy to obtain u_m for $m \geq 1$; at the Mth order, we have

$$u(x,t) = \sum_{m=0}^{M} u_m(x,t) \tag{2.46}$$

When $M \to +\infty$, we obtain an accurate approximation of the original Equation 2.35.

In the present analysis, we calculate the critical radius and flux distribution in a finite cylinder having height h and radius a, as shown in Figure 2.3a.

The time-independent diffusion equation in the nuclear reactor dynamics is given by

$$\nabla^2 \phi(r,z) + B^2 \phi(r,z) = 0 \tag{2.47}$$

Here, the buckling of reactors B^2 is given by

$$B^2 = \frac{\nu\Sigma_f - \Sigma_a}{D} \tag{2.48}$$

where
   ν is the average number of neutrons emitted per fission
   Σ_f is the macroscopic fission cross section
   Σ_a is the macroscopic absorption cross section
   D is the diffusion coefficient

Using Laplacian in cylindrical coordinates, we obtain

$$\frac{1}{r}\frac{d}{dr}\left(r\frac{d\phi(r,z)}{dr}\right) + \frac{d^2\phi(r,z)}{dz^2} + B^2\phi(r,z) = 0 \tag{2.49}$$

Then for applying the separation of variable method, let us consider the following equation:

$$\phi(r,z) = R(r)Z(z) \tag{2.50}$$

Consequently, the two separated differential equations are as follows:

$$\frac{1}{rR}\frac{d}{dr}\left(r\frac{dR}{dr}\right) + B^2 = \alpha^2 \tag{2.51}$$

$$\frac{1}{Z}\frac{d^2Z}{dz^2} = -\alpha^2 \tag{2.52}$$

where $\alpha^2 \ (>0)$ is the separation constant.
   For solving the radial part of the finite cylinder, we can write Equation 2.51 as

$$\frac{1}{r}\frac{d}{dr}\left(r\frac{dR}{dr}\right) + B_\alpha^2 R = 0 \tag{2.53}$$

where

$$B_\alpha^2 + \alpha^2 = B^2 \tag{2.54}$$

Replacing R with ϕ and considering

$$x = B_\alpha r \tag{2.55}$$

we obtain

$$x^2\phi''(x) + x\phi'(x) + x^2\phi(x) = 0$$

This implies

$$\phi''(x) + \frac{1}{x}\phi'(x) + \phi(x) = 0 \qquad (2.56)$$

Now, we will apply the HAM in Equation 2.56 by taking an initial approximation:

$$\phi_0(x) = C \qquad (2.57)$$

where C is a constant.

The nth-order deformation for Equation 2.56 is obtained as

$$L\left[\phi_n(x) - \chi_n\phi_{n-1}(x)\right] = hH(x)\Re_n\left[\vec{\phi}_{n-1}(x)\right] \qquad (2.58)$$

where:

$$\Re_n\left(\vec{\phi}_{n-1}\right) = \frac{1}{(n-1)!} \frac{\partial^{n-1}N\left[\phi(x;q)\right]}{\partial q^{n-1}}\bigg|_{q=0} \qquad (2.59)$$

$$N\left[\phi(x;q)\right] = \phi''(x;q) + \left(\frac{1}{x}\right)\phi'(x;q) + \phi(x;q) \text{ and } \phi(x;q) = \phi_0(x) + \sum_{m=1}^{\infty}\phi_m(x)q^m$$

$$\phi_m(x) = \frac{1}{m!} \frac{\partial^m\phi(x;q)}{\partial q^m}\bigg|_{q=0}$$

Now, the solution of the first deformation of Equation 2.56, by considering $h = -1$ and $H(x,t) = 1$, is given by

$$\phi_1(x) = -L^{-1}\left[\phi_0''(x) + \frac{1}{x}\phi_0'(x) + \phi_0(x)\right], \quad \text{where } \chi_1 = 0$$

Without loss of generality, we define the differential operator

$$L \equiv \frac{1}{x}\frac{d}{dx}\left(x\frac{d}{dx}\right) \qquad (2.60)$$

Here, we choose the inverse of the differential operator $L^{-1} \equiv \int_0^x (dx/x)\int_0^x x(\cdot)dx$

Therefore, we obtain the first deformation equation as

$$\phi_1(x) = -\int_0^x \frac{dx}{x}\int_0^x x\left[\phi_0(x)\right]dx$$

$$= \frac{-Cx^2}{4} \qquad (2.61)$$

Similarly, the solution for second deformation equation is

$$\phi_2(x) = \phi_1(x) - L^{-1}\left[\phi_1''(x) + \frac{1}{x}\phi_1'(x) + \phi_1(x)\right], \quad \text{where } \chi_2 = 1$$

$$= \frac{-Cx^2}{4} - \int_0^x \frac{dx}{x}\int_0^x x\left[\phi_1''(x) + \frac{1}{x}\phi_1'(x) + \phi_1(x)\right]dx \qquad (2.62)$$

$$= \frac{Cx^4}{64}$$

The solution for third deformation equation is

$$\phi_3(x) = \phi_2(x) - L^{-1}\left[\phi_2''(x) + \frac{1}{x}\phi_2'(x) + \phi_2(x)\right]$$

$$= \frac{Cx^4}{64} - \int_0^x \frac{dx}{x}\int_0^x x\left[\phi_2''(x) + \frac{1}{x}\phi_2'(x) + \phi_2(x)\right]dx \qquad (2.63)$$

$$= \frac{-Cx^6}{64\times 6\times 6}$$

Hence, we finally obtain the solution for the radial part of finite cylinder as

$$\phi(x) = \sum_{k=0}^\infty C\frac{(-1)^k}{4^k k!k!}(x)^{2k} \qquad (2.64)$$

Equation 2.64 can be written as

$$R_\alpha(\beta_\alpha r) = \sum_{k=0}^\infty C\frac{(-1)^k}{4^k k!k!}(\beta_\alpha r)^{2k}, \quad \text{where } x = \beta_\alpha r \text{ and } \phi = R_\alpha \qquad (2.65)$$

Similarly, the solution for axial part, which is given in Equation 2.52, is obtained as

$$Z_\alpha(z) = \cos(\alpha z) \qquad (2.66)$$

Thus, the final solution of the time-independent diffusion equation (Equation 2.47) is

$$\phi(r,z) = \sum_{n=0}^\infty A_n R_n(\beta_\alpha r) Z_\alpha(z) \qquad (2.67)$$

The numerical results are provided for one-speed fast neutrons in [235]U. The results reveal that the HAM provides an accurate alternative to the Legendre function–based solutions for cylindrical reactor geometry [35]. This also holds when HAM is applied for the fixed source NDE; in this case the HAM provides results in a rather straightforward manner compared to those of the separation of variables approach, which involves tedious algebraic manipulations of complicated mathematical expressions.

2.6.2 Numerical Results for Cylindrical Reactor

To provide numerical results, 1 MeV neutrons diffusing in pure ^{235}U are considered. The following data will be used [36,37].

$$v = 2.42, \sigma_f = 1.336b, \sigma_s = 5.959b, \sigma_c = 0.153b,$$

$$N_c = 0.0478 \times 10^{24} \text{ atoms cm}^{-3}$$

The diffusion coefficient $D = 1/(3N_c\sigma_{total})$ is equal to 0.9363 cm, where $\sigma_{total} = \sigma_f + \sigma_s + \sigma_c$.

Here, the absorption coefficient is $\sigma_a = \sigma_f + \sigma_c$ and the corresponding macroscopic absorption cross section is $\Sigma_a = N_c\sigma_a$. Similarly, the macroscopic fission cross section is $\Sigma_f = N_c\sigma_f$.

In order to calculate the solution for the flux distribution, the above data will be used.

2.6.3 Calculation for Critical Radius of Cylinder

We consider the ZF boundary conditions (or flux is zero at the boundary, i.e., the advective and diffusive fluxes must exactly balance) for obtaining the critical radius for a finite cylinder. According to ZF boundary conditions, for the flux to vanish on the top and bottom surfaces of the cylinder

$$\phi\left(r, \pm\frac{h_c}{2}\right) = 0 \tag{2.68}$$

Now, we apply the boundary conditions on the axial part:

$$Z_\alpha\left(\pm\frac{h_c}{2}\right) = 0 \tag{2.69}$$

or

$$\cos\left(\pm\alpha\frac{h_c}{2}\right) = 0 \tag{2.70}$$

Moreover, Equation 2.70 yields

$$\alpha\frac{h_c}{2} = n\frac{\pi}{2}, \quad \text{where } n = 1, 2, 3, \cdots \tag{2.71}$$

Then, the corresponding eigenvalues are

$$\alpha = \frac{m\pi}{h_c}, \quad m = 1, 2, 3, \cdots \tag{2.72}$$

For the first zero of the solution of the axial part, that is, $m = 1$, we write

$$Z(z) = \cos\left(\frac{\pi}{h_c} z\right) \tag{2.73}$$

Next, we apply the boundary condition on the radial part:

$$R_\alpha\left(\beta_\alpha a_c\right) = 0 \tag{2.74}$$

From Equation 2.65, by calculating the first zero, we obtain $\beta_\alpha a_c = 2.40483$.

Using the values of α and β_α, the critical radius depends upon critical heights (Table 2.2). Then, buckling B^2 for a reactor from Equation 2.54 is

$$B^2 = \left(\frac{2.40483}{a_c}\right)^2 + \left(\frac{\pi}{h_c}\right)^2 \tag{2.75}$$

The flux distribution in the finite cylinder is given by

$$\phi(r, z) = C\left[\sum_{p=0}^{\infty} \frac{(-1)^p}{4^p p!p!}\left(\frac{2.40483}{a_c} r\right)^{2p}\right]\cos\left(\frac{\pi z}{h_c}\right) \tag{2.76}$$

where C is the normalization constant. The maximized flux occurs at the center of the cylinder ($r = 0$, $z = 0$) and the flux decreases when it goes toward any surface. Finally, the flux is zero at $r = a_{c,ZF}$ and $z = H_c/2$. Table 2.2 represents

TABLE 2.2

Critical Radii for Corresponding Critical Heights

h_c	a_c at ZF	a_c at EBC
15	11.3144	9.44178
20	9.4788	7.6062
25	8.88551	7.01291
30	8.60656	6.73396
35	8.45055	6.57795
40	8.35371	6.48111
45	8.28922	6.41662
50	8.24399	6.37139
55	8.21100	6.3384
60	8.18617	6.31357

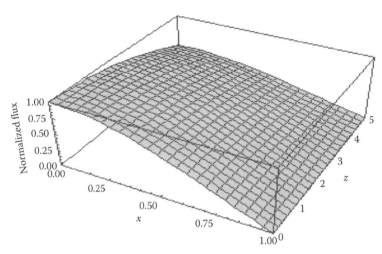

FIGURE 2.4
Flux distribution of a finite circular cylinder.

the calculated critical radii of a finite cylinder for a set of chosen critical height values. The smallest critical height is $h_c = 15$ cm and the corresponding critical radius is $a_c = 11.3144$ and $a_c = 9.44178$ at extrapolated boundary condition (EBC). The flux distributions for a finite cylinder are shown in Figure 2.4.

2.6.4 Solution for Bare Hemisphere Using ADM

The ADM is applied to formulate a new analytic solution of the NDE for a hemisphere. Different boundary conditions are investigated, including zero flux on boundary, zero flux on extrapolated boundary, and radiation boundary condition (RBC). Numerical results are provided for one-speed fast neutrons in ^{235}U. A comparison with Bessel function–based solutions demonstrates that the ADM can exactly reproduce the results more easily and efficiently. Let us consider the hemisphere as shown in Figure 2.3b.

The time-independent diffusion equation is

$$D\nabla^2\phi(r,\theta)+(v\Sigma_f-\Sigma_a)\phi(r,\theta)=0 \tag{2.77}$$

Considering the Laplacian in spherical coordinates and using $\mu=\cos\theta$, Equation 2.77 reduces to

$$\frac{\partial^2\phi(r,\mu)}{\partial r^2}+\frac{2}{r}\frac{\partial\phi(r,\mu)}{\partial r}+\frac{1}{r^2}\frac{\partial}{\partial\mu}\left[(1-\mu^2)\frac{\partial\phi(r,\mu)}{\partial\mu}\right]+B^2\phi(r,\mu)=0 \tag{2.78}$$

where the buckling of reactors B^2 is given by

$$B^2=\frac{v\Sigma_f-\Sigma_a}{D} \tag{2.79}$$

Now, by applying the separation of variables method, let us consider the following equation: $\phi(r,\mu) = R(r)\Phi(\mu)$.

Consequently, we get the differential equations as follows:

$$\frac{r^2}{R(r)}\frac{d^2R(r)}{dr^2} + \frac{2r}{R(r)}\frac{dR(r)}{dr} + B^2r^2 = n(n+1) \tag{2.80}$$

$$\frac{1}{\Phi(\mu)}\frac{d}{d\mu}\left[(1-\mu^2)\frac{d\Phi(\mu)}{d\mu}\right] = n(n+1) \tag{2.81}$$

For solving the radial part, we apply ADM [3,38]. To illustrate ADM, let us consider the general form of a differential equation:

$$Fy = g \tag{2.82}$$

where F is the nonlinear differential operator with linear and nonlinear terms. The differential operator is decomposed as

$$F \equiv L + R \tag{2.83}$$

where:

L is the easily invertible linear operator

R is the remainder of the linear operator

For our convenience, L is taken as the highest-order derivative; then Equation 2.82 can be written as

$$Ly + Ry + Ny = g \tag{2.84}$$

where Ny corresponds to the nonlinear term. Solving Ly from the above equation, we have

$$Ly = g - Ry - Ny \tag{2.85}$$

Because L is invertible, L^{-1} is the integral operator:

$$L^{-1}(Ly) = L^{-1}(g) - L^{-1}(Ry) - L^{-1}(Ny) \tag{2.86}$$

If L is a second-order operator, L^{-1} is a twofold integral operator:

$$L^{-1} \equiv \int_0^t \int_0^t (\cdot)\,dt\,dt \quad \text{and} \quad L^{-1}(Ly) = y(t) - y(0) - ty'(0)$$

Then, Equation 2.86 for y yields

$$y = y(0) + ty'(0) + L^{-1}(g) - L^{-1}(Ry) - L^{-1}(Ny) \tag{2.87}$$

Let us consider the unknown function $y(t)$ in the infinite series as

$$y(t) = \sum_{n=0}^{\infty} y_n \qquad (2.88)$$

The nonlinear term $N(y)$ will be decomposed by the infinite series of Adomian polynomials A_n $(n \geq 0)$:

$$Ny = \sum_{n=0}^{\infty} A_n \qquad (2.89)$$

where A_n's are obtained by

$$A_n = \frac{1}{n!}\left[\frac{d^n}{d\lambda^n}N\left(\sum_{i=0}^{\infty} y_i\lambda^i\right)\right]_{\lambda=0} \qquad (2.90)$$

Now, substituting Equations 2.88 and 2.89 into Equation 2.87, we obtain

$$\sum_{n=0}^{\infty} y_n = y(0) + ty'(0) + L^{-1}(g) - L^{-1}\left[R\left(\sum_{n=0}^{\infty} y_n\right)\right] - L^{-1}\left[\left(\sum_{n=0}^{\infty} A_n\right)\right] \qquad (2.91)$$

Consequently we can obtain

$$y_0 = y(0) + ty'(0) + L^{-1}(g)$$
$$y_1 = -L^{-1}R(y_0) - L^{-1}(A_0)$$
$$y_2 = -L^{-1}R(y_1) - L^{-1}(A_1) \qquad (2.92)$$
$$\vdots$$
$$y_{n+1} = -L^{-1}R(y_n) - L^{-1}(A_n)$$

and so on.

Based on the ADM, we shall consider the solution $y(t)$ as

$$y \cong \sum_{k=0}^{n-1} y_k = \varphi_n \quad \text{with} \quad \lim_{n\to\infty}\varphi_n = y(t)$$

We can apply this method to many real physical problems, and the obtained results are of high accuracy. In most of the physical problems, the practical solution φ_n the n-term approximation is convergent and accurate even for small values of n.

Here, we first consider $x = Br$ and rewrite Equation 2.80 as

$$x^2 \frac{d^2 R(x)}{dx^2} + 2x \frac{dR(x)}{dx} + \left[x^2 - n(n+1) \right] R(x) = 0 \tag{2.93}$$

$$\frac{d^2 R(x)}{dx^2} + \frac{2}{x} \frac{dR(x)}{dx} - \frac{n(n+1)}{x^2} R(x) = -R(x) \tag{2.94}$$

We compare Equation 2.94 with a general second-order differential equation:

$$f''(x) + a(x)f'(x) + b(x)f(x) = h(x) \tag{2.95}$$

Let $\varphi(x) \neq 0$ be a solution for the corresponding homogeneous differential equation of Equation 2.95.

By applying the method of variation of parameters, we obtain the general solution of Equation 2.95 as

$$f(x) = C_1 \varphi(x) + C_2 \varphi(x) \int \frac{dx}{E(x)\varphi^2(x)} + \varphi(x) \int \frac{1}{E(x)\varphi^2(x)} \left[\int E(x)\varphi(x)h(x)dx \right] dx \tag{2.96}$$

where:
$$E(x) = e^{\int a(x)dx}$$
C_1 and C_2 are constants

In the above form, L^{-1} is defined as an indefinite integral; for any other problem, we have to transform it into a definite integral according to the solution for our problem.

We consider the Bessel equation in the following form:

$$f''(x) + \frac{1}{x} f'(x) - \frac{v^2}{x^2} f(x) = -f(x) \tag{2.97}$$

The solution for the homogeneous part is $\varphi(x) = x^{\pm v}$; considering $C_1 = 2^{-v}/\Gamma(v+1)$, $C_2 = 0$, we obtain the solution of Equation 2.97 according to Equation 2.96 in the form of integral equation, which is given by

$$f(x) = \frac{(x/2)^v}{\Gamma(v+1)} - x^v \int_0^x \left[x^{-1-2v} \int_0^x x^{1+v} f(x)dx \right] dx \tag{2.98}$$

The initial approximation, which is the solution of homogeneous part of Equation 2.97 is

$$f_0(x) = \frac{(x/2)^v}{\Gamma(v+1)} \tag{2.99}$$

From above we get the next approximation as

$$f_1(x) = -x^v \int_0^x \left[x^{-1-2v} \int_0^x x^{1+v} f_0(x) dx \right] dx$$

$$= -\frac{(x/2)^{v+2}}{\Gamma(v+2)}$$

Continuing in this manner, we can find the solution in the series form given by

$$f(x) = \sum_{k=0}^{\infty} (-1)^k \frac{(x/2)^{v+2k}}{k!\Gamma(v+k+1)} \tag{2.100}$$

For this bare hemisphere, we consider $\phi(x) = x^n$ to be a solution for the corresponding homogeneous equation (Equation 2.93).

By applying the method of variation of parameters and with the help of the integral equation (Equation 2.96), we obtain the general solution for Equation 2.94 as

$$R(x) = Cx^n - x^n \int_0^x \frac{1}{x^2 x^{2n}} \left[\int_0^x x^2 x^n R(x) dx \right] dx \tag{2.101}$$

where:
$E(x) = e^{\int (2/x) dx} = x^2$
$C_1 = C$ is a constant
$C_2 = 0$

The initial approximation is

$$R_0(x) = Cx^n \tag{2.102}$$

Next, the first iterative value is

$$R_1(x) = -x^n \int_0^x \frac{1}{x^{2n+2}} \left[\int_0^x x^{n+2} R_0(x) dx \right] dx$$

$$= -Cx^n \int_0^x \frac{1}{x^{2n+2}} \left[\int_0^x x^{n+2} x^n(x) dx \right] dx \tag{2.103}$$

$$= \frac{-Cx^{n+2}}{(4n+6)}$$

The second iterative value is

$$R_2(x) = -x^n \int_0^x \frac{1}{x^{2n+2}} \left[\int_0^x x^{n+2} R_1(x)dx \right] dx$$

$$= \frac{Cx^n}{4n+6} \int_0^x \frac{1}{x^{2n+2}} \left[\int_0^x \frac{x^{2n+5}}{(2n+5)} dx \right] dx \qquad (2.104)$$

$$= \frac{Cx^{n+4}}{(4n+6)(8n+20)}$$

In general, the solution for the differential equation (Equation 2.93) is

$$R_n(Br) = \sum_{k=0}^{\infty} C_n \frac{(-1)^k \Gamma(2n+2)\Gamma(n+k+1)}{\Gamma(n+1)\Gamma(k+1)\Gamma(2n+2k+1)} (Br)^{n+2k} \qquad (2.105)$$

while the solution for angular part, namely, Equation 2.81, is just the Legendre polynomial:

$$\Phi_n(\mu) = P_n(\mu) = \sum_{k=0}^{n} \frac{1}{2^k k!} \frac{d^k}{d\mu^k} (\mu^2 - 1)^k \qquad (2.106)$$

The final solution of Equation 2.78 is therefore

$$\phi(r,\mu) = \sum_{n=0}^{\infty} C_n R_n(Br) P_n(\mu) \qquad (2.107)$$

2.6.5 Numerical Results for Hemispherical Symmetry

In order to provide numerical results, we adopt the same data as taken in Section 2.6.2. According to the ZF boundary condition, for the flux to vanish on the angular part (on the flat surface), $\mu = 0$; the properties from Legendre's polynomial implies that the even amplitudes must be equal to zero.

On the other hand, applying the ZF boundary condition on the radial part

$$R_n(Br) = 0$$

The first zero calculated using Equation 2.105 is $Ba_c = 4.4934$.

This reduces the summation in Equation 2.107 to the following simple form:

$$\phi(r,\mu) = C_1 R_1(Br) P_1(\mu) = \sum_{k=0}^{\infty} C_1 \frac{(-1)^k (6)(k+1)}{\Gamma(2k+4)} (Br)^{1+2k} \mu \qquad (2.108)$$

Hence, we obtain, for a hemisphere, the critical radius is $a_{c,ZF} = 15.085$ cm. Since the flux is assumed to converge to zero at $a_c + 2D$ using the EBC, $a_{c,EBC} = 13.185$ cm.

2.6.6 Radiation Boundary Condition

After applying the two simple boundary conditions, namely, ZF and EBC, the system yields inaccurate results. Now, we will apply the RBC, which gave more accurate results for the critical radius. We consider the two hemispheres of radius a separated by distance $2b$ from their flat surfaces. The condition can be written in the form as [39,40]

$$n.\nabla\phi = g(r)\phi \tag{2.109}$$

where $g(r)$ varies over surface; the condition on the curved surface is

$$\left.\frac{\partial\phi(r,\mu)}{\partial r}\right|_{r=a} = -g_1\phi(a,\mu) \tag{2.110}$$

For the flat surface where $\theta = \pi/2$, the condition becomes

$$\left.\frac{1}{r}\frac{\partial\phi(r,\mu)}{\partial\mu}\right|_{\mu=0} = g_2\phi(r,0) \tag{2.111}$$

By Marshak's P1 boundary condition [39]

$$g_1 = \frac{1}{2D} \tag{2.112}$$

$$g_2 = \frac{g(b/a)}{2D} \tag{2.113}$$

where $g(x)$ is given by the following equation [37,38]:

$$g(x) = \frac{1 - 2\int_0^1 d\mu\mu \, \exp\left[-f(x)\left(\sqrt{1-\mu^2}/\mu\right)\right]}{1 + 3\int_0^1 d\mu\mu^2 \, \exp\left[-f(x)\left(\sqrt{1-\mu^2}/\mu\right)\right]} \tag{2.114}$$

with

$$f(x) = \frac{x}{1+x}\left(\frac{2}{\pi} + \sqrt{2}x\right) \tag{2.115}$$

TABLE 2.3

Critical Radius for Hemisphere in Two Different Boundary Conditions

BC	ADM	Cassell and Williams [36]	Khasawneh, Dababneh, and Odibat HPM [40]
ZF	15.085	15.06	15.085
EBC	13.185	13.19	13.185
RBC	11.804	11.80397	11.804

For a sphere $b = 0$ and hence $x = 0$, which causes g and g_2 to vanish; but in the hemisphere, b and x tend to infinity, where g becomes unity (Table 2.3).

In order to solve $\phi(r,\mu)$ we apply RBC on the flat surface:

$$\frac{1}{r}\sum A_n R_n(Br)P_n'(0) = g_2 \sum A_n R_n(Br)P_n(0) \tag{2.116}$$

By the properties of Legendre polynomials:

$$P_n'(0) = 0, \quad n \text{ is even}$$

$$P_n(0) = 0, \quad n \text{ is odd}$$

$$P_{2n}(0) = \frac{(-1)^n \Gamma(n+1/2)}{\sqrt{\pi}n!}, \quad n = 0,1,2,\cdots$$

$$P_{2n+1}'(0) = \frac{2(-1)^n \Gamma(n+3/2)}{\sqrt{\pi}n!}, \quad n = 0,1,2,\cdots$$

Equation 2.116 reduces to

$$\frac{1}{r}\sum A_{2n+1}R_{2n+1}(Br)\frac{2(-1)^n \Gamma(n+3/2)}{\sqrt{\pi}n!} = g_2 \sum A_{2n}R_{2n}(Br)\frac{(-1)^n \Gamma(n+1/2)}{\sqrt{\pi}n!} \tag{2.117}$$

Using the recurrence relation

$$\sum_{n=0}^{\infty} R_{2n}\frac{(-1)^n \Gamma(n+1/2)}{\sqrt{\pi}n!}\left[2B\left(n+\frac{1}{2}\right)A_{2n+1} - g_2 A_{2n}\right]$$

$$+ \sum_{n=0}^{\infty} A_{2n-1}(2nB)\frac{R_{2n}}{(4n+1)(4n-1)}\frac{(-1)^n \Gamma(n+1/2)}{\sqrt{\pi}n!} = 0 \tag{2.118}$$

A_n, the amplitudes, are related by

$$B(2n+1)A_{2n+1} - \frac{2nB}{(4n+1)(4n-1)}A_{2n-1} = g_2 A_{2n} \tag{2.119}$$

The odd-order amplitudes can be written in terms of even one by

$$A_{2n+1} = \sum_{k=0}^{n} \gamma_{nk} A_{2k} \tag{2.120}$$

where:

$$\gamma_{nk} = \frac{(2n)!!(2k-1)!!(4k+1)!!}{(2k)!!(2n+1)!!(4n+1)!!} \tag{2.121}$$

Similarly, by applying RBC on a curved surface

$$\sum_{n=0}^{\infty} A_n R_n'(Ba) P_n(\mu) = -g_1 \sum_{n=0}^{\infty} A_n R_n(Ba) P_n(\mu) \tag{2.122}$$

or

$$\sum_{n=0}^{\infty} A_n f_n P_n(\mu) = 0 \tag{2.123}$$

where:

$$f_n = \left(\frac{n}{a} + g_1\right) R_n(Ba) - \frac{B}{(2n+3)} R_{2n+1}(Ba) \tag{2.124}$$

To make computational implementation, we introduce the shifted Legendre polynomial:

$$P_n^*(\mu) = P_n(2\mu - 1) \tag{2.125}$$

which form a complete orthogonal set over $0 < \mu < 1$, while the classical Legendre polynomial $P_n(\mu)$ forms an orthogonal set over $-1 < \mu < 1$:

$$P_n(\mu) = \sum_{k=0}^{n} \beta_{nk} P_n^*(\mu) \tag{2.126}$$

where:

$$\beta_{nk} = (2k+1) \int_0^1 d\mu P_m(\mu) P_k(2\mu - 1) \tag{2.127}$$

Hence, Equation 2.123 becomes

$$\sum_{n=0}^{\infty} A_n f_n \sum_{k=0}^{n} \beta_{nk} P_k^*(\mu) = 0 \tag{2.128}$$

By interchanging the order of summation, $\beta_{nk} = 0$ for $n < k$; we obtain

$$\sum_{k=0}^{\infty} \left(\sum_{n=0}^{\infty} \beta_{nk} f_n A_n \right) P_k^*(\mu) = 0 \qquad (2.129)$$

which yields

$$\sum_{n=0}^{\infty} \beta_{nk} f_n A_n = 0, \quad k = 0,1,2,... \qquad (2.130)$$

Now we separate the even and odd amplitudes and with use of Equation 2.120, we get

$$\sum_{k=0}^{\infty} \hat{H}_{nk} A_{2n} = 0, \quad k = 0,1,2,\cdots \qquad (2.131)$$

where:

$$\hat{H}_{nk} = \beta_{2n,k} f_{2n} + \sum_{l=n}^{\infty} \beta_{2l+1,k} f_{2l+1} \gamma_{ln} \qquad (2.132)$$

In order to get the numerical results by the cutoff value N, the infinite sum becomes

$$\sum_{n=0}^{N} \hat{H}_{nk} A_{2n} = 0, \quad k = 0,1,2,...,N \qquad (2.133)$$

where:

$$\hat{H}_{nk} = \beta_{2n,k} f_{2n} + \sum_{l=n}^{N} \beta_{2l+1,k} f_{2l+1} \gamma_{ln} \qquad (2.134)$$

Therefore, the series in the general solution contains $2N + 2$ terms:

$$\phi(r,\mu) = \sum_{n=0}^{2N+1} A_n R_n(Br) P_n(\mu) \qquad (2.135)$$

In order to calculate the amplitudes A_{2n} by Equation 2.133, we obtain

$$\sum_{n=1}^{N} \hat{H}_{nk} A_{2n} = -\hat{H}_{0k} A_0 \qquad (2.136)$$

where:

$$A_1 = 1 \text{ and } A_0 = 1/\gamma_{00} \qquad (2.137)$$

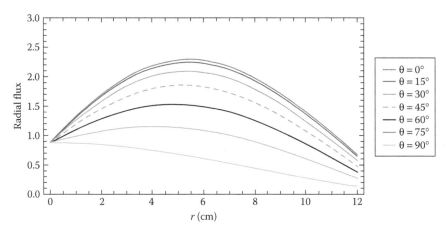

FIGURE 2.5
Flux distribution across a bare hemisphere along different angles.

The calculated flux is normalized to the volume-averaged flux $\bar{\phi}$:

$$\bar{\phi} = \frac{3}{a^3} \int\limits_0^a dr\, r^2 \int\limits_0^1 d\mu\, \phi(r,\mu) \tag{2.138}$$

where $a = 11.80396$ cm and using $N = 22$, the value of volume-averaged flux is 0.633771.

The numerical solution for flux distribution of a hemisphere at different angles is shown graphically in Figure 2.5.

Using the HAM and ADM, we successfully obtain the approximate solution of neutron flux for the one-group diffusion equation [36] in the closed form of infinite convergent series in two symmetrical bodies.

2.7 Application of the ADM for One-Group NDEs

The purpose of this study is to present the application of an analytical and approximate method for the solution of the one-group NDE, when delayed neutrons are averaged by one group of delayed neutrons [22]:

$$\frac{1}{v_c} \frac{\partial^\alpha \varphi(x,t)}{\partial t^\alpha} = D\nabla^2 \varphi(x,t) + \left(\gamma \Sigma_f - \Sigma_a\right)\varphi(x,t) + \lambda C(x,t)$$

$$\frac{\partial^\alpha C(x,t)}{\partial t^\alpha} = \beta\gamma\Sigma_f \varphi(x,t) - \lambda C(x,t) \tag{2.139}$$

where:

$0 < \alpha < (1/2)$ and with the initial conditions $\varphi(x,0) = \varphi_0(x)$

v_c is the neutron velocity

$\varphi = \varphi(x,t)$ is the neutron flux

$C(x,t)$ is the density of precursors

D is the neutron diffusion coefficient

γ is the average number of neutrons produced per fission

Σ_f is the macroscopic fission cross section

Σ_a is the macroscopic absorption cross section

λ is the radioactive decay constant

β is the fractional of the fission neutrons that are delayed

2.7.1 Solution of the Problem by the ADM

Rewriting Equation 2.139 by replacing $B = \beta\gamma\Sigma_f$ and $\Sigma = \gamma\Sigma_f - \Sigma_a$, we get

$$\frac{1}{v_c}\frac{\partial^\alpha \varphi(x,t)}{\partial t^\alpha} = D\nabla^2\varphi(x,t) + \Sigma\varphi(x,t) + \lambda C(x,t)$$

(2.140)

$$\frac{\partial^\alpha C(x,t)}{\partial t^\alpha} = B\varphi(x,t) - \lambda C(x,t)$$

Now from the first equation of Equation 2.140 we get

$$C(x,t) = \frac{1}{\lambda v_c}\frac{\partial^\alpha \varphi(x,t)}{\partial t^\alpha} - \frac{D}{\lambda}\nabla^2\varphi(x,t) - \frac{\Sigma}{\lambda}\varphi(x,t)$$

Introducing $C(x,t)$ in the second equation of Equation 2.140 we get

$$\frac{\partial^{2\alpha} \varphi(x,t)}{\partial t^{2\alpha}} = Dv_c\frac{\partial^\alpha\left[\nabla^2\varphi(x,t)\right]}{\partial t^\alpha} + \Sigma v_c\frac{\partial^\alpha \varphi(x,t)}{\partial t^\alpha} - \lambda\frac{\partial^\alpha \varphi(x,t)}{\partial t^\alpha} \qquad (2.141)$$

$$+\lambda Bv_c\varphi(x,t) + \lambda Dv_c\nabla^2\varphi(x,t) + \lambda v_c\Sigma\varphi(x,t)$$

where $0 < \alpha < (1/2)$.

In this analysis, a coupled fractional differential equation has been transformed into one fractional differential equation and then an analytical approximate method (ADM) has been applied to obtain the solution of this transformed fractional differential equation.

Applying $J_t^{2\alpha}$ on both sides of Equation 2.141, it becomes

$$\varphi(x,t) = \varphi_0(x) + J_t^{2\alpha}D_t^\alpha\left[Dv_c\nabla^2\varphi(x,t) + \Sigma v_c\varphi(x,t) - \lambda\varphi(x,t)\right]$$

(2.142)

$$+\lambda Bv_c J_t^{2\alpha}\varphi(x,t) + D\lambda v_c J_t^{2\alpha}\nabla^2\varphi(x,t) + \lambda v_c\Sigma J_t^{2\alpha}\varphi(x,t)$$

where:

$J_t^\beta f \equiv [1/\Gamma(\beta)]\int_0^t (t-\tau)^{\beta-1} f(\tau)\,d\tau$, $\beta > 0$ is Riemann–Liouville fractional integral of order β

$D_t^\beta f \equiv [1/\Gamma(k-\beta)](d^k/dt^k)\int_0^t [f(\tau)/(t-\tau)^{\beta+1-k}]$, $k-1 \leq \beta < k$ is the Riemann–Liouville fractional derivative of order β

$$\text{Let } \varphi(x,t) = \sum_{i=0}^{\infty} \varphi_i(x,t) \tag{2.143}$$

Solving Equation 2.142 by ADM we get the following.

- *First iteration*:

$$\varphi_0(x,t) = \varphi_0(x) \tag{2.144}$$

- *Second iteration*:

$$\varphi_1(x,t) = M_1(x)\frac{t^\alpha}{\Gamma(\alpha+1)} + M_2(x)\frac{t^{2\alpha}}{\Gamma(2\alpha+1)} \tag{2.145}$$

where:

$$M_1(x) = Dv_c\nabla^2\varphi_0(x) + \Sigma v_c\varphi_0(x) - \lambda\varphi_0(x)$$

$$M_2(x) = \lambda Bv_c\varphi_0(x) + D\lambda v_c\nabla^2\varphi_0(x) + \lambda v_c\Sigma\varphi_0(x)$$

- *Third iteration*:

$$\varphi_2(x,t) = N_1(x)\frac{t^{2\alpha}}{\Gamma(2\alpha+1)} + N_2(x)\frac{t^{3\alpha}}{\Gamma(3\alpha+1)} + N_3(x)\frac{t^{4\alpha}}{\Gamma(4\alpha+1)} \tag{2.146}$$

where:

$$N_1(x) = Dv_c\nabla^2 M_1(x) + \Sigma v_c M_1(x) - \lambda M_1(x)$$

$$N_2(x) = Dv_c\nabla^2 M_2(x) + \Sigma v_c M_2(x) - \lambda M_2(x) + \lambda Bv_c M_1(x)$$

$$+ D\lambda v_c\nabla^2 M_1(x) + \lambda v_c\Sigma M_1(x)$$

$$N_3(x) = \lambda Bv_c M_2(x) + D\lambda v_c\nabla^2 M_2(x) + \lambda v_c\Sigma M_2(x)$$

- *Fourth iteration*:

$$\varphi_3(x,t) = G_1(x)\frac{t^{3\alpha}}{\Gamma(3\alpha+1)} + G_2(x)\frac{t^{4\alpha}}{\Gamma(4\alpha+1)} + G_3(x)\frac{t^{5\alpha}}{\Gamma(5\alpha+1)}$$

$$+ G_4(x)\frac{t^{6\alpha}}{\Gamma(6\alpha+1)} \tag{2.147}$$

where:

$$G_1(x) = Dv_c\nabla^2 N_1(x) + \Sigma v_c N_1(x) - \lambda N_1(x)$$

$$G_2(x) = Dv_c\nabla^2 N_2(x) + \Sigma v_c N_2(x) - \lambda N_2(x) + \lambda Bv_c N_1(x) + D\lambda v_c\nabla^2 N_1(x)$$
$$+ \lambda v_c \Sigma N_1(x)$$

$$G_3(x) = Dv_c\nabla^2 N_3(x) + \Sigma v_c N_3(x) - \lambda N_3(x) + \lambda Bv_c N_2(x) + D\lambda v_c\nabla^2 N_2(x)$$
$$+ \lambda v_c \Sigma N_2(x)$$

$$G_4(x) = \lambda Bv_c N_3(x) + D\lambda v_c\nabla^2 N_3(x) + \lambda v_c \Sigma N_3(x)$$

and so on.

Therefore, the four-term solution of Equation 2.140 is

$$\varphi(x,t) = \varphi_0(x,t) + \varphi_1(x,t) + \varphi_2(x,t) + \varphi_3(x,t) \tag{2.148}$$

2.7.2 Numerical Results and Discussions

In this analysis, we use the solution (Equation 2.148) to present the nature of the neutron flux for different values of fractional order α.

The values $v_c = 220,000$ cm/s, $B = 0.000735$ cm^{-1}, $D = 0.356$ cm, $\lambda = 0.08\,\text{s}^{-1}$, $\Sigma = 0.005$ cm^{-1}, and $\varphi(x,0) = 1.0$ are represented in Table 2.4 and Figure 2.6.

The values $v_c = 220,000$ cm/s, $B = 0.000735$ cm^{-1}, $D = 0.356$ cm, $\lambda = 0.08\,\text{s}^{-1}$, $\Sigma = 0.005$ cm^{-1} and $\varphi(x,0) = x^2$ are represented in Figure 2.7.

TABLE 2.4

Neutron Flux (φ) for Different Values of Time (t) and α

Time (t)	$\alpha = 0.1$	$\alpha = 0.2$	$\alpha = 0.3$	$\alpha = 0.4$	$\alpha = 0.5$
0.0001	1.04535×10^8	6.21599×10^6	358,370	20,098.8	1,137.43
0.00039	1.59662×10^8	1.42279×10^7	1.22266×10^6	101,416	8,236.38
0.00068	1.89937×10^8	1.99782×10^7	2.02159×10^6	197,079	18,694.4
0.00097	2.12264×10^8	2.48265×10^7	2.78898×10^6	301,545	31,632.6
0.00126	2.3039×10^8	2.91405×10^7	3.53571×10^6	412,616	46,640
0.00155	2.45852×10^8	3.3087×10^7	4.26719×10^6	529,060	63,462.9
0.00184	2.59449×10^8	3.67595×10^7	4.98668×10^6	650,071	81,923.8
0.00213	2.71652×10^8	4.02168×10^7	5.69632×10^6	775,079	101,890
0.00242	2.8277×10^8	4.34989×10^7	6.39764×10^6	903,656	123,257
0.00271	2.93014×10^8	4.66344×10^7	7.09176×10^6	1.03547×10^6	145,939
0.003	3.02536×10^8	4.96447×10^7	7.77958×10^6	1.17024×10^6	169,866

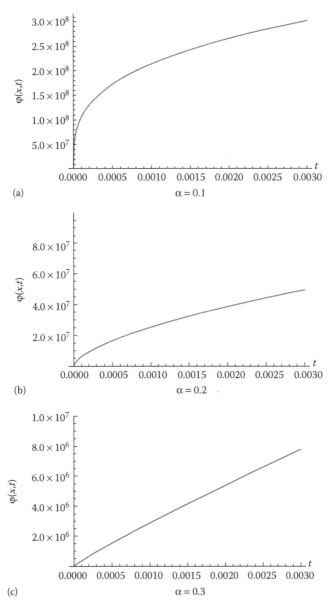

FIGURE 2.6
Two-dimensional graphs of neutron flux $\varphi(x,t)$ for $\alpha = 0.1$ (a), 0.2 (b), 0.3 (c), respectively.
(*Continued*)

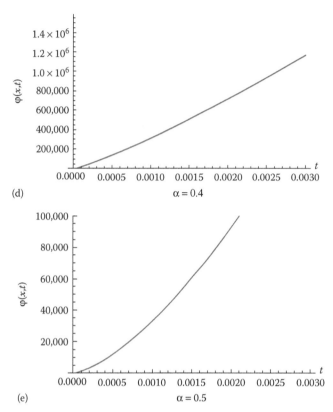

FIGURE 2.6 (Continued)
Two-dimensional graphs of neutron flux $\varphi(x, t)$ for $\alpha = 0.4$ (d), and 0.5 (e), respectively.

2.8 Conclusion

In this chapter, the VIM and MDM have been successfully applied to find the analytical approximate as well as numerical solutions of the NDE with fixed source. The decomposition method is straightforward, without restrictive assumptions, and the components of the series solution can be easily computed using any mathematical symbolic package. The NDE for symmetrical bodies like cylindrical reactors and hemispheres have been solved using the HAM and ADM. The analytical approximate as well as numerical solutions of critical radius and flux distribution for bare hemisphere and cylindrical reactor have also been obtained easily and accurately. The HAM is a powerful and efficient technique that yields analytical solutions for the one-group NDE. Using the decomposition method, it is easy to compute any kind of equation by providing more convergent series in real physical

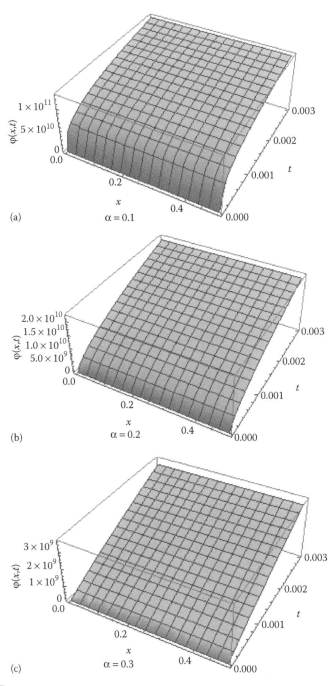

FIGURE 2.7
Three-dimensional graphs of neutron flux $\varphi(x,t)$ for $\alpha = 0.1$ (a), 0.2 (b), 0.3 (c), respectively.

(*Continued*)

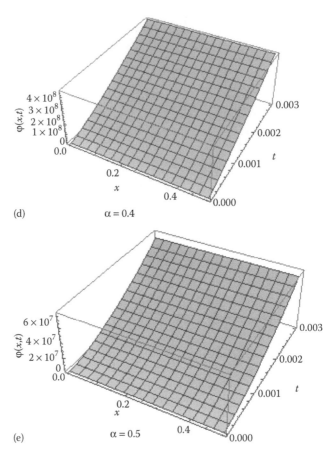

(d) $\alpha = 0.4$

(e) $\alpha = 0.5$

FIGURE 2.7 (Continued)
Three-dimensional graphs of neutron flux $\varphi(x, t)$ for $\alpha = 0.4$ (d), 0.5 (e), respectively.

problems, which can be solved with the help of any mathematical package. Moreover, neither does it require any type of discretization for variables nor does it affect computational round-off errors. The computational size of fast convergence for series solution will be reduced. We can solve many functional equations such as ordinary differential equations, partial differential equations, and integral equations using these analytical methods. It requires minimal computer memory and is free from rounding-off errors and discretization of space variables. These methods can be used to obtain solutions for diffusion and kinetic equations in the dynamic system of nuclear reactors.

Nuclear reactor operation will be safe only if we control the reactor operation. We must therefore understand the process in the reactor core. Neutron flux is one of the main reactor parameters. The aim of this chapter also includes application of the ADM for the analytical approximate solution of the fractional NDE, which describes neutron transport in nuclear reactor.

3

Fractional Order Neutron
Point Kinetic Model

3.1 Introduction

In this chapter, fractional calculus and the numerical solution for fractional neutron point kinetic equation (FNPKE) are discussed. The technique for efficient and accurate numerical computation for FNPKE with different values of reactivity is introduced. The fractional neutron point kinetic (FNPK) model has been analyzed for the dynamic behavior of neutron motion; here, the relaxation time associated with a variation in the neutron flux involves a fractional order acting as an exponent of the relaxation time, which results in optimum operation of the nuclear reactor. The results of studies on the neutron dynamic behavior for subcritical, critical, and supercritical reactivities as well as for different values of fractional order have been presented and compared with those of the classical neutron point kinetic (CNPK) model.

3.2 Brief Description for Fractional Calculus

Fractional calculus is three centuries old; however, it is not popular among the science and/or engineering community. The traditional integral and derivative are, to say the least, a staple for the technology professional, essential as a means of understanding and working with natural and artificial systems. Fractional calculus is a branch of calculus that generalizes the derivative of a function to noninteger order. The beauty of this subject is that fractional derivatives are not a local property. Fractional differential equations (FDEs) appear more and more frequently in different research areas and engineering applications. In recent years, considerable interest in FDEs has been stimulated due to their numerous applications in the areas of physics and engineering. Many important phenomena in electromagnetics, acoustics, viscoelasticity, electrochemistry, control theory, neutron point kinetic model, anomalous diffusion, Brownian motion, signal and image processing, fluid dynamics, and materials science are well described by differential equations of fractional order.

Fractional calculus is a field of applied mathematics that deals with derivatives and integrals of arbitrary orders. It is also known as the generalized integral and differential calculus of arbitrary order (Kilbas et al. [41] and Sabatier et al. [42]). Fractional calculus was described by Gorenflo and Mainardi [43] as the field of mathematical analysis that deals with investigation and applications of integrals and derivatives of arbitrary order. And many great mathematicians (in pure and applied mathematics), such as Abel, Caputo, Euler, Fourier, Grünwald, Hadamard, Hardy, Heaviside, Holmgren, Laplace, Leibniz, Letnikov, Liouville, Riemann, Riesz, and Weyl, made major contributions to the theory of fractional calculus. The history of fractional calculus dates to the end of the seventeenth century; the birth of fractional calculus was due to an exchange of letters. At that time scientific journals did not exist and scientists exchanged information through letters. The first conference on fractional calculus and its applications was organized in June 1974 by Ross at the University of New Haven, Connecticut.

In recent years, fractional calculus has become the focus of interest for many researchers in different disciplines of applied science and engineering because of the fact that a realistic modeling of a physical phenomenon can be successfully achieved by using fractional calculus. Fractional calculus has gained considerable significance during the past decades mainly due to its applications in diverse fields of science and engineering. For the purpose of this study Caputo's definition of the fractional derivative will be used, where the initial conditions for FDEs [44] with Caputo's derivatives take on the traditional form, as for integer-order differential equations. For that reason, we need a reliable and efficient technique to arrive at the solution of FDEs.

3.2.1 Definition: Riemann–Liouville Integral and Derivative Operator

The concept of nonintegral order of integration can be traced to the philosopher and creator of modern calculus G. W. Leibniz, who made some remarks on the meaning and possibility of fractional derivative of the order 1/2 in late seventeenth century. However, rigorous investigation was first carried out by Liouville; this resulted in a series of papers between 1832 and 1837, where he defined the first fractional integral. Later investigations and further developments by many others led to the construction of the integral-based Riemann–Liouville fractional integral operator, which has been a valuable cornerstone in fractional calculus ever since. Prior to Liouville and Riemann, Euler took the first step in the study of fractional integration when he studied the simple case of fractional integrals of monomials of arbitrary real order in the heuristic fashion of time. It has been said to have lead him to construct the gamma function for fractional powers of the factorial. An early attempt by Liouville was later purified by the Swedish mathematician Holmgren [45], who in 1865 made important contributions to the growing study of fractional calculus. However, it was Riemann [46] who reconstructed it to fit Abel's integral equation, and thus made it vastly more useful. Today, there exist many different forms of fractional integral operators, ranging from divided-difference types to infinite-sum types, but the Riemann–Liouville operator is still the most frequently used when fractional integration is performed.

The most frequently encountered definition of an integral of fractional order is the Riemann–Liouville integral [44], in which the fractional integral of order α (>0) is defined as

$$J^{\alpha}f(t) = \frac{1}{\Gamma(\alpha)} \int_{0}^{t} (t-\tau)^{\alpha-1} f(\tau) d\tau, \quad t > 0, \alpha \in R^{+} \tag{3.1}$$

where:

R^{+} is the set of positive real numbers

The gamma function Γ is defined by $\Gamma(n) = \int_{0}^{\infty} e^{-t} t^{n-1} dt$ and for real number n, $\Gamma(n) = 1, 2, 3, \ldots (n-1)$.

The Riemann–Liouville fractional derivative is defined by

$$D^{\alpha} f(t) = D^{m} J^{m-\alpha} f(t) = \frac{d^{m}}{dt^{m}} \left[\frac{1}{\Gamma(m-\alpha)} \int_{0}^{t} \frac{f(\tau)}{(t-\tau)^{\alpha-m+1}} d\tau \right], \tag{3.2}$$

$$m - 1 < \alpha < m, m \in N$$

Left—Riemann–Liouville fractional derivative can be defined by

$$_{a}D_{t}^{\alpha} f(t) = \frac{1}{\Gamma(n-\alpha)} \left(\frac{d}{dt} \right)^{n} \int_{a}^{t} (t-\tau)^{n-\alpha-1} f(\tau) d\tau \tag{3.3}$$

Right—Riemann–Liouville fractional derivative can be defined by

$$_{t}D_{b}^{\alpha} f(t) = \frac{1}{\Gamma(n-\alpha)} \left(-\frac{d}{dt} \right)^{n} \int_{t}^{b} (\tau-t)^{n-\alpha-1} f(\tau) d\tau \tag{3.4}$$

Fractional Riemann–Liouville derivatives have various interesting properties. For example, the fractional derivative of a constant is not zero, namely

$$_{a}D_{t}^{\alpha} C = \frac{C(t-a)^{-\alpha}}{\Gamma(1-\alpha)}, \quad \text{where } 0 < \alpha < 1, \alpha \in R^{+} \tag{3.5}$$

3.2.2 Definition: Caputo's Fractional Derivative

There is another option for computing fractional derivatives, the Caputo fractional derivative. It was introduced by M. Caputo in his 1967 research paper. In contrast to the Riemann–Liouville fractional derivative, when solving differential equations using Caputo's definition [43,44], it is not necessary to define the fractional order initial conditions. Caputo's definition is illustrated as follows:

$$D_t^\alpha f(t) = J^{m-\alpha} D^m f(t)$$

$$= \begin{cases} \dfrac{1}{\Gamma(m-\alpha)} \displaystyle\int_0^t (t-\tau)^{(m-\alpha-1)} \dfrac{d^m f(\tau)}{d\tau^m} d\tau, & \text{if } m-1 < \alpha < m, \ m \in N \quad (3.6) \\[4mm] \dfrac{d^m f(t)}{dt^m}, & \text{if } \alpha = m, \ m \in N \quad (3.7) \end{cases}$$

where:
   the parameter α is the order of the derivative and is allowed to be real or
   even complex

For the Caputo's derivative, we have

$$D^\alpha C = 0 \tag{3.8}$$

where:
   C is a constant

$$D^\alpha t^\beta = \begin{cases} 0, & \beta \le \alpha - 1 \\[3mm] \dfrac{\Gamma(\beta+1)t^{\beta-\alpha}}{\Gamma(\beta-\alpha+1)}, & \beta > \alpha - 1 \end{cases} \tag{3.9}$$

Similar to integer-order differentiation, Caputo's derivative is linear.

$$D^\alpha \left[\gamma f(t) + \delta g(t) \right] = \gamma D^\alpha f(t) + \delta D^\alpha g(t) \tag{3.10}$$

where:
   γ and δ are constants, and satisfies the so-called Leibniz's rule

$$D^\alpha \left[g(t)f(t) \right] = \sum_{k=0}^{\infty} \binom{\alpha}{k} g^{(k)}(t) D^{\alpha-k} f(t) \tag{3.11}$$

if $f(\tau)$ is continuous in $[0, t]$ and $g(\tau)$ has continuous derivatives sufficient
number of times in $[0, t]$.

3.2.3 Grünwald–Letnikov Definition of Fractional Derivatives

In mathematics, the Grünwald–Letnikov (GL) fractional order derivative
is a basic extension of the derivative in fractional calculus that allows one
to take the derivative a noninteger number of times. It was introduced by
Anton Karl Grünwald (1838–1920) from Prague, Czech Republic, in 1867,
and by Aleksey Vasilievich Letnikov (1837–1888) from Moscow, in 1868. GL
fractional derivative [41–44] is defined by

$$_aD_t^p f(t) = \lim_{\substack{h \to 0 \\ nh=t-a}} h^{-p} \sum_{r=0}^{n} \omega_r^p f(t-rh)$$ (3.12)

where:

$$\omega_r^p = (-1)^r \binom{p}{r}$$

$$\omega_0^p = 1 \text{ and } \omega_r^p = \left(1 - \frac{p+1}{r}\right)\omega_{r-1}^p, r = 1, 2, \cdots$$

Lemma

If $m - 1 < \alpha < m$, $m \in N$, then

$$D^\alpha J^\alpha f(t) = f(t)$$ (3.13)

and

$$J^\alpha D^\alpha f(t) = f(t) - \sum_{k=0}^{m-1} \frac{t^k}{k!} f^{(k)}(0+), \quad t > 0$$ (3.14)

In the field of nuclear engineering, the neutron diffusion and point kinetic equations are the most vital models; they have been included to countless studies and applications under neutron dynamics and its effects. With the help of the neutron diffusion concept, we understand the complex behavior of average neutron motion. From many reactor studies, we get the idea that neutron motion is a diffusion process. It is also assumed that the neutrons are in average motion diffused at very low or high neutron density. The concept of neutron transport as a diffusion process has only limited validation due to that neutron stream at relatively large distances between interactions. The process of neutron diffusion takes place in a very highly heterogeneous hierarchical configuration. Here, we propose a numerical scheme for the solution of fractional diffusion model as a constitutive equation of neutron current density. FNPKE has proved particularly useful in the context of anomalous slow diffusion.

3.3 FNPKE and Its Derivation

The diffusion theory model of neutron transport has played a crucial role in reactor theory since it is simple enough to allow scientific insight, and it is sufficiently realistic to study many important design problems. The mathematical methods used to analyze such a model are the same as those applied in more sophisticated methods such as multigroup neutron transport theory. The neutron flux (ψ) and current (J) in the diffusion theory model are related in a simple way under certain conditions. This relationship between ψ and J is identical in form to a law used in the study of diffusion phenomena in liquids and gases, namely, Fick's law. The use of this law in reactor theory leads to the diffusion approximation, which is a result of a number of simplifying assumptions. On the other hand, higher-order neutron transport codes have always been deployed in nuclear engineering for mostly time-independent problems out-of-core shielding calculations.

FNPK model is based on the diffusion theory using all known theoretical arguments. The scope of the FNPK is to describe the neutron transient behavior in a highly heterogeneous configuration in nuclear reactors, in the presence of strong neutron absorbers in the fuel, control rods, and chemical shim in the coolant. In summary, there are many interesting problems to consider from the point of view of the FDEs: the challenge is the modeling and simulation of the new generation of nuclear reactors, as well as the advanced molten salt reactor [47], where the old paradigms can no longer be valid.

The fractional model retains the main dynamic characteristics of the neutron motion. The physical interpretation of the fractional order is related with non-Fickian effects from the neutron diffusion equation point of view.

To derive the FNPKE [48–50] for point reactor, we consider with a source term given by

$$\frac{\tau^\kappa}{v} \frac{\partial^{\kappa+1}\phi}{\partial t^{\kappa+1}} + \tau^\kappa \sum_a \frac{\partial^\kappa \phi}{\partial t^\kappa} + \frac{1}{v}\frac{\partial \phi}{\partial t} = S - \sum_a \phi + D\nabla^2\phi + \tau^\kappa \frac{\partial^\kappa S}{\partial t^\kappa} \qquad (3.15)$$

where:
  ϕ is the neutron flux
  S is the source term
  v is the average number of neutrons emitted per fission
  D is the diffusion coefficient
  Σ_a is the macroscopic absorption cross section

The one-group source term S [23] is given by

$$S = (1-\beta)k_\infty \sum_a \phi + \sum_{i=1}^m \lambda_i \hat{C}_i \qquad (3.16)$$

where:
  k_∞ is the effective multiplication factor
  $\phi = \phi(r,t)$; $S = S(r,t)$; and $\hat{C}_i = \hat{C}_i(r,t)$ are all functions of position and time
  λ_i is the precursor decay constant for group i

By substituting the above equation in Equation 3.15, we obtain

$$\frac{\tau^{\kappa}}{v}\frac{\partial^{\kappa+1}\phi}{\partial t^{\kappa+1}}+\tau^{\kappa}\left[\sum_{a}+(1-\beta)k_{\infty}\sum_{a}\right]\frac{\partial^{\kappa}\phi}{\partial t^{\kappa}}+\frac{1}{v}\frac{\partial\phi}{\partial t}$$

$$=\left[(1-\beta)k_{\infty}\sum_{a}-\sum_{a}\right]\phi+DV^{2}\phi+\sum_{i=1}^{m}\lambda_{i}\hat{C}_{i}+\tau^{\kappa}\sum_{i=1}^{m}\left(\lambda_{i}\frac{\partial^{\kappa}}{\partial t^{\kappa}}\hat{C}_{i}\right) \tag{3.17}$$

According to Glasstone and Sesonske (1981), $\nabla^{2}\phi$ is replaced by $-B_{g}^{2}\phi$, where B_{g}^{2} is geometric buckling. Then, Equation 3.17 can be written as

$$\frac{\tau^{\kappa}}{v}\frac{\partial^{\kappa+1}\phi}{\partial t^{\kappa+1}}+\tau^{\kappa}\left[\sum_{a}+(1-\beta)k_{\infty}\sum_{a}\right]\frac{\partial^{\kappa}\phi}{\partial t^{\kappa}}+\frac{1}{v}\frac{\partial\phi}{\partial t}$$

$$=\left[(1-\beta)k_{\infty}\sum_{a}-\sum_{a}-DB_{g}^{2}\right]\phi+\sum_{i=1}^{m}\lambda_{i}\hat{C}_{i}+\tau^{\kappa}\sum_{i=1}^{m}\left(\lambda_{i}\frac{\partial^{\kappa}}{\partial t^{\kappa}}\hat{C}_{i}\right) \tag{3.18}$$

Now by using $\phi=vn(t)$, Equation 3.18 leads to

$$\tau^{\kappa}\frac{d^{\kappa+1}n(t)}{dt^{\kappa+1}}+\tau^{\kappa}\left[\sum_{a}+(1-\beta)k_{\infty}\sum_{a}\right]v\frac{d^{\kappa}n(t)}{dt^{\kappa}}+\frac{dn(t)}{dt}$$

$$=\left[(1-\beta)k_{\infty}\sum_{a}-\sum_{a}-DB_{g}^{2}\right]vn(t)+\sum_{i=1}^{m}\lambda_{i}\hat{C}_{i}+\tau^{\kappa}\sum_{i=1}^{m}\left(\lambda_{i}\frac{\partial^{\kappa}}{\partial t^{\kappa}}\hat{C}_{i}\right) \tag{3.19}$$

In order to solve the FNPKE, we consider the following definition for nuclear parameters [23,51]:

- Diffusion area: $L^{2}=D/\Sigma_{a}$
- Prompt neutron lifetime: $l=1/v\Sigma_{a}(1+L^{2}B_{g}^{2})$
- Neutron generation time: $\Lambda=1/k_{\infty}v\Sigma_{a}$
- Effective multiplication factor: $k_{\text{eff}}=k_{\infty}/(1+L^{2}B_{g}^{2})$
- Reactivity: $\rho=(k_{\text{eff}}-1)/k_{\text{eff}}$

The fractional equation for NPKE [49] is given by

$$\tau^{\kappa}\frac{d^{\kappa+1}n}{dt^{\kappa+1}}+\tau^{\kappa}\left[\frac{1}{l}+\frac{(1-\beta)}{\Lambda}\right]\frac{d^{\kappa}n}{dt^{\kappa}}+\frac{dn}{dt}=\left(\frac{\rho-\beta}{\Lambda}\right)n+\sum_{i=1}^{m}C_{i}\lambda_{i}+\tau^{\kappa}\sum_{i=1}^{m}\left(\lambda_{i}\frac{d^{\kappa}C_{i}}{dt^{\kappa}}\right) \tag{3.20}$$

where:
τ is the relaxation time
κ is the anomalous diffusion order (for subdiffusion process: $0 < k < 1$, while for super diffusion process: $1 < k < 2$)
n is the neutron density
C_{i} is the concentration of the ith neutron delayed precursor
l is the prompt-neutron lifetime for infinite media
β is the fraction of delayed neutrons

When $\tau^\kappa \to 0$, the CNPK equation can be obtained as

$$\frac{dn}{dt} = \left(\frac{\rho-\beta}{\Lambda}\right)n + \sum_{i=1}^{m} C_i\lambda_i \qquad (3.21)$$

The net rate of formation of the precursor of delayed neutrons corresponding to the ith group is given by

$$\frac{dC_i}{dt} = \left(\frac{\beta_i}{\Lambda}\right)n - \lambda_i C_i \qquad (3.22)$$

Equations 3.20 and 3.22 describe the neutron dynamics process in the nuclear reactor and the delayed neutrons precursor of the ith group, respectively.

The numerical approximation of the solution of the FNPK model is obtained by applying the numerical method like explicit finite difference method [13,52].

Considering one group of delayed neutrons, the FNPKE and initial conditions are given by

$$\tau^\kappa \frac{d^{\kappa+1}n}{dt^{\kappa+1}} + \tau^\kappa\left[\frac{1}{l} + \frac{(1-\beta)}{\Lambda}\right]\frac{d^\kappa n}{dt^\kappa} + \frac{dn}{dt} = \left(\frac{\rho-\beta}{\Lambda}\right)n + \lambda C + \lambda\tau^\kappa \frac{d^\kappa C}{dt^\kappa}, \qquad (3.23)$$

$$\text{where } 0 < \kappa \le 2$$

with

$$n(0) = n_0 \qquad (3.24)$$

The precursor concentration balance equation is

$$\frac{dC}{dt} = \frac{\beta n}{\Lambda} - \lambda C \qquad (3.25)$$

with

$$C(0) = C_0 = \frac{\beta}{\lambda\Lambda}n_0 \qquad (3.26)$$

The initial conditions for fractional neutron model are presented by

$$n(0) = n_0 = 1$$

$$\left.\frac{d^{\kappa+1}n}{dt^{\kappa+1}}\right|_{t=0} = 0 \qquad (3.27)$$

$$\left.\frac{d^\kappa n}{dt^\kappa}\right|_{t=0} = 0$$

In order to analyze the effect of anomalous diffusion order (κ) and relaxation time (τ) on the behavior of the neutron density, the numerical model was implemented for the solution of kinetic equation through the simulation in a computer program.

The nuclear parameters used were obtained from [24,53]

$$\beta = \sum_{i=1}^{6} \beta_i = 0.007$$

$$\lambda = \frac{\beta}{\sum_{i=1}^{6} \beta_i/\lambda_i} = 0.0810958 \text{ s}^{-1}$$

where β_i and λ_i are presented in Table 3.1, the parameter value for l was obtained from Kinard and Allen [53]

$$\Lambda = 0.002 \text{ s}$$

$$l = 0.00024 \text{ s}$$

$$\tau^{\kappa} = 10^{-4} \text{s}^{\kappa}$$

where:

$$C_0 = 43.1588$$

$$n_0 = 1$$

A series of numerical computations are carried out by using explicit finite difference in order to obtain a best approximation for nuclear dynamics of FNPK model. We considered four cases of anomalous diffusion order (κ), namely, $\kappa = 0.99, 0.98, 0.97,$ and 0.96, respectively, for relaxation time $\tau^{\kappa} = 10^{-4} \text{s}^{\kappa}$ ($s \equiv 1$ second).

TABLE 3.1

Neutron Delayed Fractions and Decay Constants

ith Group	β_i	λ_i
Group 1	0.000266	0.0127
Group 2	0.001491	0.0317
Group 3	0.001316	0.1550
Group 4	0.002849	0.3110
Group 5	0.0008960	1.4000
Group 6	0.000182	3.8700

Source: Kinard, M., Allen, K.E.J., *Ann. Nucl. Energ.*, 31, 1039–1051, 2004.

3.4 Application of Explicit Finite Difference Scheme for FNPKE

For the fractional model Equation 3.23 through 3.26, let us take the time step size h. Using the definition of GL fractional derivative, the numerical approximation of the Equation 3.23 through 3.26, in view of the research work [13,52], is

$$\tau^\kappa h^{-(\kappa+1)} \sum_{j=0}^{m} \omega_j^{\kappa+1} n_{m-j} + \tau^\kappa \left[\frac{1}{l} + \frac{(1-\beta)}{\Lambda} \right] h^{-\kappa} \sum_{j=0}^{m} \omega_j^\kappa n_{m-j} + h^{-1}[n_m - n_{m-1}]$$

$$= \left(\frac{\rho - \beta}{\Lambda} \right) n_m + \lambda C_m + \lambda \tau^\kappa h^{-\kappa} \sum_{j=0}^{m} \omega_j^\kappa C_{m-j}$$

(3.28)

For the precursor concentration balance equation, the scheme is

$$h^{-1}[C_m - C_{m-1}] = \frac{\beta n_m}{\Lambda} - \lambda C_m$$

(3.29)

where:
$n_m = n(t_m)$ is the mth approximation for neutron density at time t_m
$C_m = C(t_m)$ is the mth approximation for precursor concentration at time t_m

Here, $n(0) = n_0 = 1$; $C_0 = \dfrac{\beta n_0}{\lambda \Lambda}$; $t_m = mh$; $m = 0,1,2,\ldots$; $\omega_j^\kappa = (-1)^j \dbinom{\kappa}{j}$; and $j = 0,1,2,3,\ldots$.

Now, Equations 3.28 and 3.29 can be written, respectively, as

$$n_m - n_{m-1} = h \left(\frac{\rho - \beta}{\Lambda} \right) n_m + \lambda h C_m + \lambda \tau^\kappa h^{1-\kappa} \sum_{j=0}^{m} \omega_j^\kappa C_{m-j}$$

$$- \tau^\kappa h^{-\kappa} \sum_{j=0}^{m} \omega_j^{\kappa+1} n_{m-j} - \tau^\kappa \left[\frac{1}{l} + \frac{(1-\beta)}{\Lambda} \right] h^{1-\kappa} \sum_{j=0}^{m} \omega_j^\kappa n_{m-j}$$

(3.30)

$$C_m - C_{m-1} = \frac{h \beta n_m}{\Lambda} - \lambda h C_m$$

The above equation leads to implicit numerical iteration scheme. However, in this work, we propose an explicit numerical scheme that leads from the time layer t_{m-1} to t_m as follows:

$$n_m = n_{m-1} + h \left(\frac{\rho - \beta}{\Lambda} \right) n_{m-1} + \lambda h C_{m-1} + \lambda \tau^\kappa h^{1-\kappa} C_{m-1}$$

$$+ \lambda \tau^\kappa h^{1-\kappa} \sum_{j=1}^{m} \omega_j^\kappa C_{m-j} - \tau^\kappa h^{-\kappa} n_{m-1} - \tau^\kappa h^{-\kappa} - \tau^\kappa \left[\frac{1}{l} + \frac{(1-\beta)}{\Lambda} \right] h^{1-\kappa} n_{m-1}$$

(3.31)

$$- \tau^\kappa \left[\frac{1}{l} + \frac{(1-\beta)}{\Lambda} \right] h^{1-\kappa} \sum_{j=1}^{m} \omega_j^\kappa n_{m-j}, \quad \text{where } m = 1,2,3,\cdots$$

Precursor concentration balance equation is

$$C_m = C_{m-1} + \frac{h\beta n_{m-1}}{\Lambda} - \lambda h C_{m-1}, \quad \text{where } m = 1, 2, 3, \cdots \tag{3.32}$$

3.5 Analysis for Stability of Numerical Computation

The stability of the numerical computation is calculated by taking the different time step size h with different values of the anomalous diffusion order κ and the relaxation time τ. In order to obtain a stable result, a set of time step sizes are considered by trial and error for different values of κ and τ^κ.

Many numerical experiments have been done for getting a better solution for FNPK. In the present numerical study, we consider $\kappa = 0.99, 0.98, 0.97$, and 0.96, respectively, for relaxation time $\tau^\kappa = 10^{-4} s^\kappa$. The values of time step size h are taken between $0.01 \le h \le 0.05$. The simulation time was considered as 1 s. The stability criterion in this analysis is related with 1% of the relative error to $n_0 = 1$ at 1 s for simulation time.

$$1\% \ge \left| \frac{n_f - n_0}{n_f} \right| \times 100$$

where:
n_f is the neutron density that is calculated from fractional model

To exhibit the behavior of neutron density with $\tau^\kappa = 10^{-4} s^\kappa$ and $\kappa = 0.99$, we consider h in the interval [0.01 s, 0.05 s]. By stability criteria, the relative error is 0.40% at $h = 0.01$ s, 0.402% at $h = 0.025$ s, and 0.405% at $h = 0.05$ s. To examine the behavior of neutron density with $\tau^\kappa = 10^{-4} s^\kappa$ and $\kappa = 0.98$, we consider h in the interval [0.01 s, 0.05 s] where the curves are very similar with $\kappa = 0.99$. In this case, the relative error is 0.39% at $h = 0.01$ s, 0.399% at $h = 0.025$ s, and 0.401% at $h = 0.05$ s. Thus, for numerical computation, we can increase the time step size higher with respect to $\kappa = 0.99$. It can be shown that the neutron density behavior with $\tau^\kappa = 10^{-4} s^\kappa$ and $\kappa = 0.97$, considering h in the interval [0.01 s, 0.05 s], the curves are very similar to $\kappa = 0.99$ and 0.98, respectively. For the case $\kappa = 0.97$, the numerical scheme is also stable for time step size $h = 0.01$ s whose corresponding relative error is 0.39%. For the case when $\kappa = 0.96$ and h in the interval [0.01 s, 0.05 s], the relative error is 0.399% at $h = 0.01$ s, 0.392% at $h = 0.025$ s, and 0.395% at $h = 0.05$ s. Therefore, the above numerical experiment confirms the stability of our proposed numerical scheme for the solution of FNPKE.

3.6 Numerical Experiments with Change of Reactivity

The behavior of the fractional model with reactivity changes [49] is illustrated graphically in this section. The neutron density is analyzed by taking different values for anomalous diffusion order [49]. In the present analysis, we have taken the time step sizes $h = 0.01, 0.025$, and 0.05. The numerical experiments involve the insertion of three reactivity steps:

- *Case I*: $\rho = 0.003$ (positive reactivity for supercritical reactor)
- *Case II*: $\rho = 0$ (reactivity for critical reactor)
- *Case III*: $\rho = -0.003$ (negative reactivity for subcritical reactor)

The fractional model was compared with a classical solution obtained by [22–24,32]

$$n(t) = n_0 \left\{ \frac{\beta}{\beta - \rho} e^{[\lambda \rho t/(\beta - \rho)]} - \frac{\rho}{\beta - \rho} e^{[-(\beta - \rho)t/\Lambda]} \right\}$$

The anomalous diffusion order considered for this numerical scheme is $\kappa = 0.99, 0.98, 0.97$, and 0.96 for relaxation time $\tau^\kappa = 10^{-4} \text{s}^\kappa$. The relaxation time increases when κ increases.

3.6.1 Case I: Results of Positive Reactivity for Supercritical Reactor

The following figure shows numerical comparison of neutron density for fractional and classical neutron point kinetic equation in case of supercritical reactor (Figures 3.1 through 3.3).

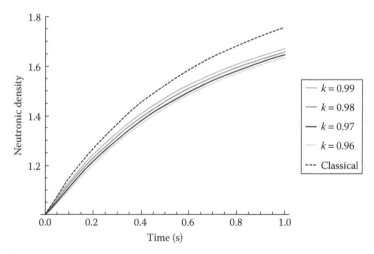

FIGURE 3.1
Comparison of neutron density behavior for FNPK and CNPK for $\rho = 0.003$, $h = 0.01$, and $\tau^\kappa = 10^{-4} \text{s}^\kappa$ with $\kappa = 0.99, 0.98, 0.97$, and 0.96, respectively.

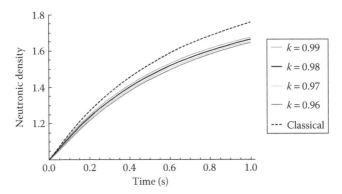

FIGURE 3.2
Comparison of neutron density behavior for FNPK and CNPK for $\rho = 0.003$, $h = 0.025$, and $\tau^{\kappa} = 10^{-4}s^{\kappa}$ with $\kappa = 0.99$, 0.98, 0.97, and 0.96, respectively.

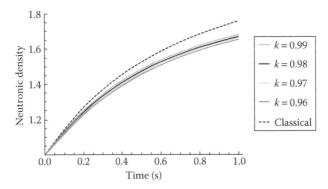

FIGURE 3.3
Comparison of neutron density behavior for FNPK and CNPK for $\rho = 0.003$, $h = 0.05$, and $\tau^{\kappa} = 10^{-4}s^{\kappa}$ with $\kappa = 0.99$, 0.98, 0.97, and 0.96, respectively.

3.6.2 Case II: Results of Reactivity for Critical Reactor

The following figure shows numerical comparison of neutron density for fractional and classical neutron point kinetic equation in case of critical reactor (Figures 3.4 through 3.6).

3.6.3 Case III: Results of Negative Reactivity for Subcritical Reactor

The following figure shows numerical comparison of neutron density for fractional and classical neutron point kinetic equation in case of subcritical reactor (Figures 3.7 through 3.9).

These results exhibit the neutron dynamic behavior for positive reactivity for supercritical reactor as shown in Figures 3.1 through 3.3, for critical reactor in Figures 3.4 through 3.6, and for negative reactivity for subcritical

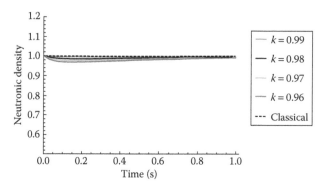

FIGURE 3.4
Comparison of neutron density behavior for FNPK and CNPK for $\rho = 0$, $h = 0.01$, and $\tau^\kappa = 10^{-4} s^\kappa$ with $\kappa = 0.99$, 0.98, 0.97, and 0.96, respectively.

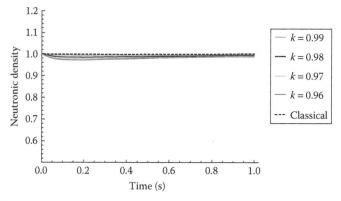

FIGURE 3.5
Comparison of neutron density behavior for FNPK and CNPK for $\rho = 0$, $h = 0.025$, and $\tau^\kappa = 10^{-4} s^\kappa$ with $\kappa = 0.99$, 0.98, 0.97, and 0.96, respectively.

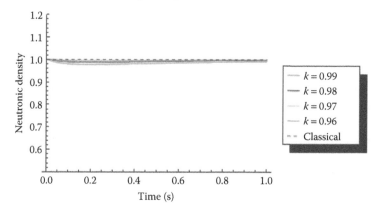

FIGURE 3.6
Comparison of neutron density behavior for FNPK and CNPK for $\rho = 0$, $h = 0.05$, and $\tau^\kappa = 10^{-4} s^\kappa$ with $\kappa = 0.99$, 0.98, 0.97, and 0.96, respectively.

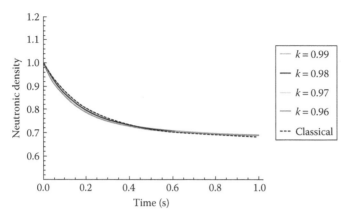

FIGURE 3.7
Comparison of neutron density behavior for FNPK and CNPK for $\rho = -0.003$, $h = 0.01$, and $\tau^\kappa = 10^{-4} s^\kappa$ with $\kappa = 0.99, 0.98, 0.97$, and 0.96, respectively.

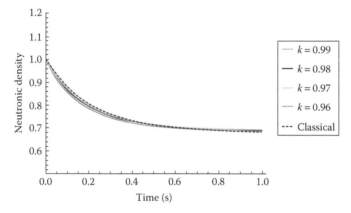

FIGURE 3.8
Comparison of neutron density behavior for FNPK and CNPK for $\rho = -0.003$, $h = 0.025$, and $\tau^\kappa = 10^{-4} s^\kappa$ with $\kappa = 0.99, 0.98, 0.97$, and 0.96, respectively.

reactor in Figures 3.7 through 3.9 for different values of time length steps and relaxation time. For critical reactor at $\rho = 0$, solution for the FNPKE coincides with classical solution. It is also compared with the CNPK model. It can be observed from the above figures for positive reactivity (Figures 3.1 through 3.3) and negative reactivity (Figures 3.7 through 3.9) that the obtained results in the present numerical scheme are in good agreement with Figure 8.18 (for positive reactivity $\rho = 0.003$) and Figure 8.21 (for negative reactivity $\rho = -0.003$) obtained in the works of Espinosa-Paredes et al. [49]. In comparison with the detrended fluctuation analysis method proposed by Espinosa-Paredes et al. [49], this explicit finite difference scheme is more convenient to use and efficient to calculate the numerical solution for FNPK model. It is

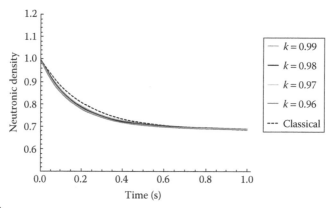

FIGURE 3.9
Comparison of neutron density behavior for fractional and classical NPK for $\rho = -0.003$, $h = 0.05$, and $\tau^\kappa = 10^{-4}\,s^\kappa$ with $\kappa = 0.99, 0.98, 0.97$, and 0.96, respectively.

easily computed by using any mathematical software package having fewer round-off errors.

Espinosa-Paredes et al. [49] have proposed a solution procedure that has been inherited from the works of Edwards et al. [54]. In order to apply numerical approximation method for the solution of FNPKE, they discretized fractional derivative using Diethelm's method [55,56]. In order to solve FNPKE, they used the numerical algorithm given by Edwards et al. [54], where the fractional kinetic model was represented as a multiterm high-order linear FDE. Then, it was converted into a system of ordinary and fractional differential equation. In contrast to their method, in the present numerical scheme fractional derivative has been discretized by GL derivative and the FNPKE has been converted directly into finite difference equation. Then, it has been adjusted in the form of explicit finite difference scheme. It is very much clear that the present numerical scheme requires less number of computational efforts compared to Espinosa-Paredes et al. [49].

3.7 Conclusion

In this chapter, the dynamics of FNPKE were studied. In this chapter, the numerical solution for one-group delayed neutron FNPKE was determined by explicit finite difference method. The three cases for change in reactivity have been discussed with respect to nuclear reactors in this point kinetic model. The numerical experiments including comparison with CNPK model were carried out for both positive and negative reactivity for different values of fractional order κ. The fractional model retains the main characteristics of neutron motion where relaxation time associated with rapid variation in neutron flux contains a fractional exponent to obtain the best approximation for nuclear reactor dynamics. The procedure of the numerical approximations to the solution of FNPK model is represented graphically using mathematical software. The numerical solutions obtained from explicit finite difference exhibit close behavior with CNPK model. It has also been observed that the obtained results in the present method are in good agreement with those obtained in the works of Espinosa-Paredes et al. [49]. This numerical method is a very efficient and convenient technique for solving FNPK model.

4

Numerical Solution for Deterministic Classical and Fractional Order Neutron Point Kinetic Model

4.1 Introduction

In the dynamical system of a nuclear reactor, the neutron point kinetic equations are the coupled linear differential equations that are used to determine neutron density and delayed neutron precursor concentrations. These kinetic equations are the most vital model in nuclear engineering. The modeling of these equations involves the use of time-dependent parameters [32]. Reactivity function and neutron source are the parametric quantities of this essential system. Neutron density and delayed neutron precursor concentrations differ randomly with respect to time. At high-power levels, random behavior is imperceptible. But at low-power levels, such as at the beginning of nuclear power generation, random fluctuations in neutron density and neutron precursor concentrations can be crucial. The proposed technique, multistep differential transform method (MDTM), that we have used in this research is based on the Taylor series expansion, which provides a solution in terms of convergent series with easily computable components. Here both classical and fractional order neutron point kinetic equations have been analyzed over step, ramp, and sinusoidal reactivity functions. Fractional calculus generates the derivative and antiderivative operations of differential and integral calculus [57–59] from noninteger orders to the entire complex plane. The semianalytical numerical technique that we applied in this research is the most transparent method available for the solution of classical as well as fractional neutron point kinetic equations.

4.2 Application of MDTM to Classical Neutron Point Kinetic Equation

In this section, we consider the classical integer order neutron point kinetic equations for m-delayed groups as follows [24,53]:

$$\frac{dn(t)}{dt} = \left[\frac{\rho(t) - \beta}{l}\right] n(t) + \sum_{i=1}^{m} \lambda_i c_i + S(t) \tag{4.1}$$

$$\frac{dc_i(t)}{dt} = \frac{\beta_i}{l} n(t) - \lambda_i c_i(t), \quad i = 1, 2, \ldots, m \tag{4.2}$$

where:
  $n(t)$ is the time-dependent neutron density
  $c_i(t)$ is the ith precursor density
  $\rho(t)$ is the time-dependent reactivity function
  β_i is the ith delayed fraction
  $\beta = \sum_{i=1}^{m} \beta_i$ is the total delayed fraction
  l is the neutron generation time
  λ_i is the ith group decay constant
  $S(t)$ is the neutron source function

The classical neutron point kinetic equation is considered in matrix form as follows [60]:

$$\frac{d\vec{x}(t)}{dt} = A\vec{x}(t) + B(t)\vec{x}(t) + \vec{S}(t) \tag{4.3}$$

where:

$$\vec{x}(t) = \begin{bmatrix} n(t) \\ c_1(t) \\ c_2(t) \\ \vdots \\ c_m(t) \end{bmatrix}$$

with initial condition $\vec{x}(0) = \vec{x}_0$
  Here, we define A as

$$A = \begin{bmatrix} \dfrac{-\beta}{l} & \lambda_1 & \lambda_2 & \cdots & \lambda_m \\ \dfrac{\beta_1}{l} & -\lambda_1 & 0 & \cdots & 0 \\ \dfrac{\beta_2}{l} & 0 & -\lambda_2 & \ddots & \vdots \\ \vdots & \vdots & \ddots & \ddots & 0 \\ \dfrac{\beta_m}{l} & 0 & \cdots & 0 & -\lambda_m \end{bmatrix}_{(m+1)\times(m+1)}$$

$B(t)$ can be expressed as

$$B(t) = \begin{bmatrix} \dfrac{\rho(t)}{l} & 0 & 0 & \cdots & 0 \\ 0 & 0 & 0 & \cdots & 0 \\ 0 & 0 & 0 & \ddots & \vdots \\ \vdots & \vdots & \ddots & \ddots & 0 \\ 0 & 0 & \cdots & 0 & 0 \end{bmatrix}_{(m+1)\times(m+1)}$$

and $\vec{S}(t)$ is defined as

$$\vec{S}(t) = \begin{bmatrix} q(t) \\ 0 \\ 0 \\ \vdots \\ 0 \end{bmatrix}$$

where:
$q(t)$ is the time-dependent neutron source term

In this section, we will apply the MDTM to obtain the solution for classical neutron point kinetic equation (Equation 4.3).

The MDTM can be described as follows:

Let us consider the following nonlinear initial value problem:

$$f(t, u, u', \ldots, u^{(p)}) = 0, \; u^{(p)} \text{ is } p \text{th derivative of } u \tag{4.4}$$

subject to the initial conditions $u^{(k)}(0) = c_k$, for $k = 0, 1, \ldots, p - 1$.

We find the solution over the interval $[0, T]$. The approximate solution of the initial value problem can be expressed by the following finite series:

$$u(t) = \sum_{m=0}^{M} a_m t^m, t \in [0, T] \tag{4.5}$$

Assume that the interval $[0, T]$ is divided into N subintervals $[t_{n-1}, t_n]$, $n = 1, 2, \ldots, N$ of equal step size $h = T/N$ by using the node point $t_n = nh$. The main idea of the MDTM is to apply first DTM to Equation 4.3 over the interval $[0, t_1]$ and obtain the following approximate solutions:

$$u_1(t) = \sum_{m=0}^{M_1} a_{1m} t^m, t \in [0, t_1] \tag{4.6}$$

using the initial conditions $u_1^{(k)}(0) = c_k$. For $n \geq 2$ and at each subinterval $[t_{n-1}, t_n]$, we use the initial conditions $u_n^{(k)}(t_{n-1}) = u_{n-1}^{(k)}(t_{n-1})$ and apply the

DTM to Equation 4.3 over the interval $[t_{n-1}, t_n]$, where t_0 in replaced by t_{n-1}. This process is repeated and generates a series of approximate solutions $u_n(t)$, $n = 1,2,...,N$. Now

$$u_n(t) = \sum_{m=0}^{M_1} a_{nm}(t - t_{n-1})^m, \ t \in [t_n, t_{n+1}] \tag{4.7}$$

where $M = M_1 \cdot N$. Hence, the MDTM assumes the following solution:

$$u(t) = \begin{cases} u_1(t), & t \in [0, t_1] \\ u_2(t), & t \in [t_1, t_2] \\ \vdots \\ u_N(t), & t \in [t_{N-1}, t_N] \end{cases} \tag{4.8}$$

The MDTM is a simple computational technique used for all values of h. It can be easily shown that if the step size $h = T$, MDTM reduces to classical DTM. The main advantage of this new algorithm is that the obtained series solution converges for wide time regions.

Here, the time domain is divided into subdomains for $i = 0,1,2,...,N$, and the approximate functions in each subdomains are $\bar{x}_i(t)$, $i = 1,2,...,N$.

By taking the differential transform method [15,61,62] of the Equation 4.3, we obtain the following:

- For step reactivity, the differential transform scheme is

$$X_i(k+1) = \frac{1}{(k+1)}[(A+B)X_i(k) + F(k)] \tag{4.9}$$

- For ramp and sinusoidal reactivity, the differential transform scheme is

$$X_i(k+1) = \frac{1}{(k+1)}\left\{ \left[AX_i(k) + \sum_{l=0}^{k} B(l)X_i(k-l) \right] + F(k) \right\} \tag{4.10}$$

where:

$$X_i(k) = \frac{1}{k!}\left\{ \frac{d^k[\bar{x}(t)]}{dt^k} \right\}\Bigg|_{t=t_i}$$

$$F(k) = \frac{1}{k!}\left\{ \frac{d^k[\vec{F}(t)]}{dt^k} \right\}\Bigg|_{t=t_i}$$

From the initial condition we can obtain $X_0(0) = \bar{x}(0) = \bar{x}_0$, accordingly from Equation 1.40 we obtain the following:

$$\vec{x}_0(t) = \sum_{r=0}^{p} (t - t_0)^r X_0(r)$$

Here,

$$\vec{x}(t_1) \cong \vec{x}_0(t_1) = \sum_{r=0}^{p} (t_1 - t_0)^r X_0(r) = \sum_{r=0}^{p} h^r X_0(r), \quad h = t_1 - t_0 \tag{4.11}$$

The final value $\vec{x}_0(t_1)$ of the first subdomain is the initial value of the second subdomain, that is, $\vec{x}_1(t_1) = X_1(0) = \vec{x}_0(t_1)$. In this manner, $\vec{x}(t_2)$ can be represented as

$$\vec{x}(t_2) \cong \vec{x}_1(t_2) = \sum_{r=0}^{p} h^r X_1(r), \quad h = t_2 - t_1 \tag{4.12}$$

Hence, the solution on the grid points t_{i+1} can be found as

$$\vec{x}(t_{i+1}) \cong \vec{x}_i(t_{i+1}) = \sum_{r=0}^{p} h^r X_i(r), \quad h = t_{i+1} - t_i \tag{4.13}$$

By using Equations 4.9 and 4.10, we can obtain the solution for constant reactivity function and time-dependent reactivity function, respectively.

4.3 Numerical Results and Discussions for Classical Neutron Point Kinetic Model Using Different Reactivity Functions

In the present analysis, we consider three cases of reactivity function: step, ramp, and sinusoidal reactivities.

4.3.1 Results Obtained for Step Reactivity

Let us consider the first example of a nuclear reactor problem [53] with $m = 6$ and neutron source free ($q = 0$) delayed group of system with the following parameters:

$$\lambda_i = [0.0127, 0.0317, 0.115, 0.311, 1.4, 3.87]$$

$$l = 0.00002, \beta = 0.007$$

$$\beta_i = [0.000266, 0.001491, 0.001316, 0.002849, 0.000896, 0.000182]$$

We consider the problem for $t \geq 0$ with three step reactivity insertions $\rho = 0.003, 0.007,$ and 0.008. Here, we assume the initial condition $\vec{x}(0)$ as

$$\vec{x}(0) = \begin{bmatrix} 1 \\ \dfrac{\beta_1}{\lambda_1 l} \\ \dfrac{\beta_2}{\lambda_2 l} \\ \vdots \\ \dfrac{\beta_m}{\lambda_m l} \end{bmatrix} \qquad (4.14)$$

The results of classical neutron point kinetic equations for different time step reactivities are presented in Tables 4.1 through 4.3. The present method is compared with those for piecewise constant approximation (PCA) [53], constant reactivity (CORE) [63], and Taylor methods [64], as well as with the exact values [65]. Also, these numerical results are illustrated in Figures 4.1 through 4.3 for the three step reactivities ρ = 0.003, 0.007, and 0.008, respectively.

4.3.2 Results Obtained for Ramp Reactivity

Now we consider a ramp reactivity of 0.01$/s and a neutron source free ($q = 0$) equilibrium system, where the following parameters are used:

TABLE 4.1

Comparison Results at Step Reactivity ρ = 0.003 for Neutron Density

Time (s)	PCA [53]	Taylor [64]	CORE [63]	MDTM	Exact [65]
$t = 1$	2.2098	2.2098	2.2098	2.20984	2.2098
$t = 10$	8.0192	8.0192	8.0192	8.0192	8.0192
$t = 20$	2.8297×10^1	2.8297×10^1	2.8297×10^1	2.82974×10^1	2.82974×10^1

TABLE 4.2

Comparison Results at Step Reactivity ρ = 0.007 for Neutron Density

Time (s)	PCA [53]	Taylor [64]	CORE [63]	MDTM	Exact [65]
$t = 0.01$	4.5088	4.5086	4.5088	4.50886	4.5088
$t = 0.5$	5.3459×10^3	5.3447×10^3	5.3458×10^3	5.34589×10^3	5.3459×10^3
$t = 2$	2.0591×10^{11}	2.0566×10^{11}	2.0600×10^{11}	2.05916×10^{11}	2.0591×10^{11}

TABLE 4.3

Comparison Results at Step Reactivity ρ = 0.008 for Neutron Density

Time (s)	PCA [53]	Taylor [64]	CORE [63]	MDTM	Exact [65]
$t = 0.01$	6.0229	6.2080	6.2029	6.20285	6.0229
$t = 0.5$	1.4104×10^3	2.1398×10^{12}	2.1071×10^{12}	2.10706×10^{12}	1.4104×10^3
$t = 2$	6.1634×10^{23}	5.6255×10^{46}	5.2735×10^{46}	5.27345×10^{46}	6.1634×10^{23}

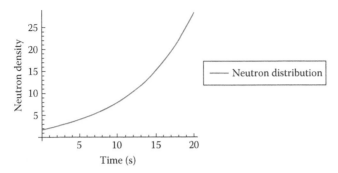

FIGURE 4.1
Neutron density at step reactivity $\rho = 0.003$.

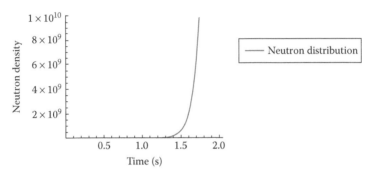

FIGURE 4.2
Neutron density at step reactivity $\rho = 0.007$.

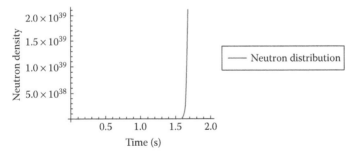

FIGURE 4.3
Neutron density at step reactivity $\rho = 0.008$.

$$\lambda_i = [0.0127, 0.0317, 0.115, 0.311, 1.4, 3.87]$$

$$\beta_i = [0.000266, 0.001491, 0.001316, 0.002849, 0.000896, 0.000182]$$

$$l = 0.00002, \; \beta = 0.007$$

with the same initial condition as given in Equation 4.14.

TABLE 4.4

Comparison Results Obtained with Ramp Reactivity for Neutron Density

Time (s)	PCA [53]	Taylor [64]	MDTM	Exact [65]
$t = 2$	1.3382	1.3382	1.3384	1.3382
$t = 4$	2.2285	2.2285	2.2287	2.2284
$t = 6$	5.5822	5.5823	5.5806	5.5821
$t = 8$	42.790	42.789	42.545	42.786
$t = 9$	487.610	487.520	471.780	487.520

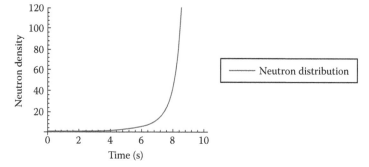

FIGURE 4.4

Neutron density for ramp reactivity calculated with the MDTM.

The ramp reactivity function is $\rho(t) = 0.1\beta t$. The results of classical neutron point kinetic equations for ramp reactivity (time-dependent function) are presented in Table 4.4 in order to exhibit the comparison results of the present method with that of the PCA method [53], the Taylor method [64], and with the exact values [65]. Also the numerical result for neutron density with ramp reactivity is illustrated in Figure 4.4.

4.3.3 Results Obtained for Sinusoidal Reactivity

Next we consider the last case, namely, sinusoidal reactivity. In this case, we consider the following kinetic parameters:

$$\lambda_i = [0.0124, 0.0305, 0.111, 0.301, 1.14, 3.01]$$

$$\beta_i = [0.000215, 0.001424, 0.001274, 0.002568, 0.000748, 0.000273]$$

$$l = 0.0005, \beta = 0.006502$$

The system is neutron source free ($q = 0$) with the same initial condition as given in Equation 4.14. The sinusoidal reactivity (time-dependent) function is $\rho(t) = \beta \sin\left(\pi t/T\right)$ where T is the half-life period ($T = 5$ s).

The comparison results between the present method (MDTM), the CORE method [63], and the Taylor method [64] are shown in Table 4.5. Also the

TABLE 4.5

Results Obtained with Sinusoidal Reactivity for Neutron Density

Time (s)	CORE [63]	Taylor [64]	MDTM
$t = 2$	10.1475	11.3820	11.3325
$t = 4$	96.7084	92.2761	90.0440
$t = 6$	16.9149	16.0317	15.5705
$t = 8$	8.8964	8.6362	8.4531
$t = 10$	13.1985	13.1987	12.9915

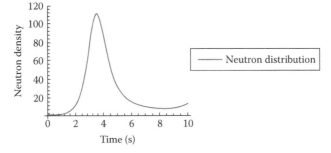

FIGURE 4.5

Neutron density for sinusoidal reactivity calculated with the MDTM.

numerical results for neutron density with sinusoidal reactivity are illustrated in Figure 4.5.

4.4 Mathematical Model for Fractional Neutron Point Kinetic Equation

Here we consider the fractional neutron point kinetic equations for m-delayed groups with Caputo derivative of order α ($\alpha > 0$ and $\alpha \in \Re$) in the field of nuclear reactor dynamics as follows:

$$\frac{d^{\alpha}\vec{x}}{dt^{\alpha}} = A\vec{x} + B(t)\vec{x} + \vec{S}(t) \tag{4.15}$$

where:

$$\vec{x}(t) = \begin{bmatrix} n(t) \\ c_1(t) \\ c_2(t) \\ \vdots \\ c_m(t) \end{bmatrix}$$

Here we define A as

$$A = \begin{bmatrix} \dfrac{-\beta}{l} & \lambda_1 & \lambda_2 & \cdots & \lambda_m \\[2mm] \dfrac{\beta_1}{l} & -\lambda_1 & 0 & \cdots & 0 \\[2mm] \dfrac{\beta_2}{l} & 0 & -\lambda_2 & \ddots & \vdots \\[2mm] \vdots & \vdots & \ddots & \ddots & 0 \\[2mm] \dfrac{\beta_m}{l} & 0 & \cdots & 0 & -\lambda_m \end{bmatrix}_{(m+1)\times(m+1)} \tag{4.16}$$

$B(t)$ can be expressed as

$$B(t) = \begin{bmatrix} \dfrac{\rho(t)}{l} & 0 & 0 & \cdots & 0 \\[2mm] 0 & 0 & 0 & \cdots & 0 \\[2mm] 0 & 0 & 0 & \ddots & \vdots \\[2mm] \vdots & \vdots & \ddots & \ddots & 0 \\[2mm] 0 & 0 & \cdots & 0 & 0 \end{bmatrix}_{(m+1)\times(m+1)} \tag{4.17}$$

and $\vec{S}(t)$ is defined as

$$\vec{S}(t) = \begin{bmatrix} q(t) \\ 0 \\ 0 \\ \vdots \\ 0 \end{bmatrix} \tag{4.18}$$

4.5 Fractional Differential Transform Method

At the beginning, we expand the analytical function $f(t)$ in terms of a fractional power series as follows:

$$f(t) = \sum_{k=0}^{\infty} F_\alpha(k)(t - t_0)^{k\alpha} \tag{4.19}$$

where $0 < \alpha \leq 2$ is the order of the fractional derivative and $F_\alpha(k)$ is the fractional differential transform of $f(t)$ given by

$$F_\alpha(k) = \frac{1}{\Gamma(\alpha k + 1)} \left\{ (D_{t_0}^\alpha)^k [f(t)] \right\}_{t=t_0} \tag{4.20}$$

TABLE 4.6

The Fundamental Operations of Fractional Differential Transform

Properties	Time Function	Fractional Transformed Function
1	$f(t) = g(t) \pm h(t)$	$F_\alpha(k) = G_\alpha(k) \pm H_\alpha(k)$
2	$f(t) = g(t)h(t)$	$F_\alpha(k) = \sum_{l=0}^{k} G_\alpha(l)H_\alpha(k-l)$
3	$f(t) = t^{p\alpha} \quad (p \in Z^+)$	$F_\alpha(k) = \dfrac{1}{\Gamma(\alpha k + 1)}\left\{(D_{t_0}^\alpha)^k\left[f(t)\right]\right\}_{t=t_0} = \begin{cases} 1, & p = k \\ 0, & p > k \\ 0, & p < k \end{cases}$ or $F_\alpha(k) = \delta(k - p)$, where $\delta(k-p) = \begin{cases} 1, k = p \\ 0, k \neq p \end{cases}$
4	$f(t) = D_{t_0}^\alpha[g(t)]$	$F_\alpha(k) = \dfrac{\Gamma\left[\alpha(k+1)+1\right]}{\Gamma(\alpha k + 1)}G_\alpha(k+1)$
5	$f(t) = \sin(\omega t + \beta)$	$F_\alpha(k) = \dfrac{1}{k!}\dfrac{d^k\left[D_{t_0}^\alpha f(t)\right]}{dt^k}\bigg|_{t=0} = \dfrac{\omega^k}{k!}\sin\left(\dfrac{\alpha k \pi}{2} + \beta\right)$ $F_\alpha(k) = \dfrac{1}{k!}\dfrac{d^k\left[D_{t_0}^\alpha f(t)\right]}{dt^k}\bigg|_{t=t_i} = \dfrac{\omega^k}{k!}\sin\left(\dfrac{\alpha k \pi}{2} + \omega t_i + \beta\right)$

where:

$(D_{t_0}^\alpha)^k = D_{t_0}^\alpha \cdot D_{t_0}^\alpha \cdot D_{t_0}^\alpha \ldots D_{t_0}^\alpha$ is the k-times-differentiable Caputo fractional derivative

Then, we can approximate the function $f(t)$ by the finite series

$$f(t) = \sum_{k=0}^{N} F_\alpha(k)(t - t_0)^{k\alpha} \tag{4.21}$$

Here, N is the finite number of terms in the truncated series solution. The basic properties of the fractional differential transform are given in Table 4.6.

4.6 Application of MDTM to Fractional Neutron Point Kinetic Equation

In this section, we will apply the DTM to obtain the solution for fractional neutron point kinetic equation (Equation 4.15). To illustrate the basic idea of the DTM for solving system of fractional differential equation, we consider the following form:

$$D^\alpha x_i(t) = N_i[x(t)], \quad i = 1, 2, \ldots, n \tag{4.22}$$

where:
$N_i[x(t)]$ is the linear or nonlinear terms of fractional differential equation
$D^\alpha x_i(t)$ is the α-order Caputo fractional derivative of unknown function $x_i(t)$

$$x_i(t) = \sum_{k=0}^{\infty} F_i(k)(t - t_0)^{k\alpha} \tag{4.23}$$

Taking the differential transform of Equation 4.23, we obtain

$$F_i(k + 1) = \frac{\Gamma(\alpha k + 1)}{\Gamma\{[\alpha(k + 1)] + 1\}} DTM\{N_i[x(t)]\}, \quad i = 1, 2, \ldots, n \tag{4.24}$$

where:

$$DTM\{N_i[x(t)]\} = \frac{1}{\Gamma(\alpha k + 1)}\left((D^\alpha)^k \{N_i[x(t)]\}\right)_{t=t_0}$$

Here, Equation 4.24 is a recursive formula with $F_i(0)$ as the value of initial condition.

We divide the interval $[0, T]$ into subintervals with time step Δt. For getting the solution in each subinterval and to satisfy the initial condition on each subinterval, the initial value $x_i(0)$ will be changed for each subinterval, that is, $x_i(t_j) = c_i^* = F_i(0)$, where $j = 0,1,2,\ldots,n - 1$. To obtain the solution on every subinterval of equal length Δt, we assume that the new initial condition is the solution in the previous interval. Thus to obtain the solution in interval $[t_j, t_{j+1}]$, the initial conditions of this interval are as follows:

$$c_i = x_i(t_j) = \sum_{m=0}^{N} F_i(m)(t_j - t_{j-1})^{m\alpha}, \quad j = 1, 2, \ldots, n \tag{4.25}$$

where:
c_i is the initial condition in the interval $[t_{j-1}, t_j]$

After applying the differential transformation [15,61,62] to Equation 4.15, we obtain the numerical scheme as follows:

- For step reactivity, the differential transform scheme is

$$X_i(k + 1) = \frac{\Gamma(\alpha k + 1)}{\Gamma\{[\alpha(k + 1)] + 1\}}[(A + B)X_i(k) + S(k)] \tag{4.26}$$

- For ramp and sinusoidal reactivities, the differential transform scheme is

$$X_i(k+1) = \frac{\Gamma(\alpha k + 1)}{\Gamma\{[\alpha(k+1)] + 1\}} \left\{ \left[AX_i(k) + \sum_{l=0}^{k} B(l)X_i(k-l) \right] + S(k) \right\} \quad (4.27)$$

where:

$$X_i(k) = \frac{1}{k!} \left(\frac{d^k\{D^{\alpha}[\vec{x}(t)]\}}{dt^k} \right) \Bigg|_{t=t_i} , \ S(k) = \frac{1}{k!} \left(\frac{d^k\{D^{\alpha}[\vec{S}(t)]\}}{dt^k} \right) \Bigg|_{t=t_i}$$

where $i = 0,1,2,\ldots,N-1$ and $k = 0,1,2,\ldots$.
  For $f(t) = t^p$, we obtain

$$F_{\alpha}(k) = \frac{1}{\Gamma(\alpha k + 1)} \{(D_{t_0}^{\alpha})^k[f(t)]\}_{t=t_0} = \begin{cases} 1, & p = k\alpha \\ 0, & p > k\alpha \\ 0, & p < k\alpha \end{cases}$$

or we can write,

$$F_{\alpha}(k) = \delta(p - k\alpha)$$

where:

$$\delta(p - k\alpha) = \begin{cases} 1, p = k\alpha \\ 0, p \neq k\alpha \end{cases}$$

From the initial condition, we can obtain that

$$X_0(0) = \vec{x}(0) = \vec{x}_0$$

where:

$$\vec{x}_0(t) = \sum_{r=0}^{p} (t - t_0)^{\alpha r} X_0(r)$$

Here,

$$c_0 \cong \vec{x}_0(t_1) = \sum_{r=0}^{p} (t_1 - t_0)^{r\alpha} X_0(r) = \sum_{r=0}^{p} h^{r\alpha} X_0(r), \ h = t_1 - t_0 \quad (4.28)$$

The final value $x_0(t_1)$ of the first subdomain is the initial value of the second subdomain, that is, $\vec{x}_1(t_1) = c_1 = \vec{x}_0(t_1)$. In this manner, $\vec{x}(t_2)$ can be represented as

$$\vec{x}(t_2) \cong \vec{x}_1(t_2) = \sum_{r=0}^{p} h^{r\alpha} X_1(r), \ h = t_2 - t_1 \tag{4.29}$$

Hence, the solution with the grid points t_{j+1} can be found as

$$\vec{x}(t_{j+1}) \cong \vec{x}_i(t_{j+1}) = \sum_{r=0}^{p} h^{r\alpha} X_i(r), \quad h = t_{j+1} - t_j \tag{4.30}$$

Using Equations 4.26 and 4.27, we can obtain the solution for the constant reactivity function and time-dependent reactivity function, respectively.

4.7 Numerical Results and Discussions for Fractional Neutron Point Kinetic Equation

In the present analysis, we consider the three cases of reactivity function such as step, ramp, and sinusoidal reactivities, respectively.

4.7.1 Results Obtained for Step Reactivity

Let us consider the first example of a nuclear reactor problem [53,64] with $m = 6$ and a neutron source free $q(t) = 0$ delayed group of a system with the following parameters:

$$\lambda_i = [0.0127, 0.0317, 0.115, 0.311, 1.4, 3.87](s^{-1}), \ l = 0.00002 \, s, \ \beta = 0.007$$

$$\beta_i = [0.000266, 0.001491, 0.001316, 0.002849, 0.000896, 0.000182]$$

We consider the problem for $t \geq 0$ with the three-step reactivity insertions $\rho = 0.003\$, 0.007\$$, and $0.008\$$, respectively. Here, we assume that the initial $\vec{x}(0)$ equals

$$\vec{x}(0) = \begin{bmatrix} 1 \\ \dfrac{\beta_1}{\lambda_1 l} \\ \dfrac{\beta_2}{\lambda_2 l} \\ \vdots \\ \dfrac{\beta_m}{\lambda_m l} \end{bmatrix} \text{ and } \vec{x}'(0) = 0 \tag{4.31}$$

The results of fractional neutron point kinetic equations are presented in Tables 4.7 through 4.9 for different time with various step reactivities. We

TABLE 4.7

Results Obtained at Subcritical Reactivity $\rho = 0.003\$$ for Neutron Density $n(t)$ Using the MDTM

t	$\alpha = 1.5$	$\alpha = 1.25$	$\alpha = 1$	$\alpha = 0.75$	McMohan and Pierson (Classical Integer Order) [64]
0.01	1.01131	1.1225	1.65208	1.80478	NA
0.5	1.39672	1.76907	1.99336	4.37265	NA
1	1.58424	1.79325	2.20988	8.9838	2.2099
10	1.78621	2.16011	8.0192	1.6548×10^6	8.0192

TABLE 4.8

Results Obtained at Critical Reactivity $\rho = 0.007\$$ for Neutron Density $n(t)$ Using the MDTM

t	$\alpha = 1.5$	$\alpha = 1.25$	$\alpha = 1$	$\alpha = 0.75$	McMohan and Pierson (Classical Integer Order) [64]
0.01	1.02659	1.31201	4.54413	51.9663	4.5086
0.5	2.31718	17.2111	5352.12	5.81118×10^{28}	5.3447×10^3
1	3.63692	38.1606	1.80784×10^6	2.10571×10^{56}	NA
2	6.29591	120.4600	2.05916×10^{11}	2.72997×10^{111}	2.0566×10^{11}

TABLE 4.9

Results Obtained at Critical Reactivity $\rho = 0.008\$$ for Neutron Density $n(t)$ Using the MDTM

t	$\alpha = 1.5$	$\alpha = 1.25$	$\alpha = 1$	$\alpha = 0.75$	McMohan and Pierson (Classical Integer Order) [64]
0.01	1.03045	1.36463	6.2694	2491.81	6.2415
0.5	2.65622	68.6659	2.11821×10^{12}	4.1051×10^{126}	6.9422×10^{11}
1	4.65868	756.7200	6.19596×10^{23}	2.20782×10^{252}	6.1215×10^{22}

compute the numerical solution taking step size $h = 0.0001$ s at time $t = 0.01$ s, 0.05 s, and 1 s, respectively, with steps = 20,000, which are also illustrated in Figures 4.6 through 4.8.

Hence, from Tables 4.7 through 4.9, it can be observed that by taking the three reactivities $\rho = 0.003\$$, $0.007\$$, and $0.008\$$, respectively, the numerical approximation results for neutron density obtained from the MDTM are in good agreement with the results obtained by McMohan and Pierson [64] using Taylor series solution for classical order $\alpha = 1$. But using these three reactivities, there are no previous results existing in the open literature for fractional order point kinetic equations.

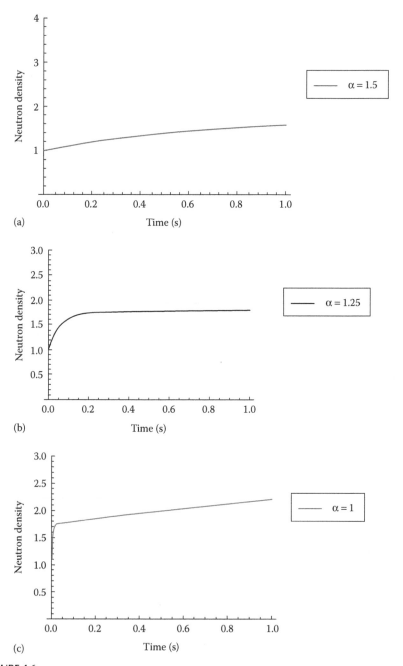

FIGURE 4.6

Neutron density for step reactivity (a) $\rho = 0.003\$$ with $\alpha = 1.5$, (b) $\rho = 0.003\$$ with $\alpha = 1.25$, and (c) $\rho = 0.003\$$ with $\alpha = 1$.   *(Continued)*

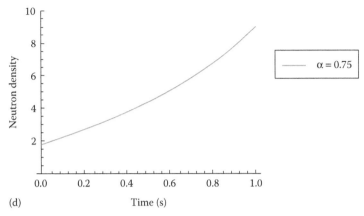

(d)

FIGURE 4.6 (Continued)
Neutron density for step reactivity (d) $\rho = 0.003\$$ with $\alpha = 0.75$.

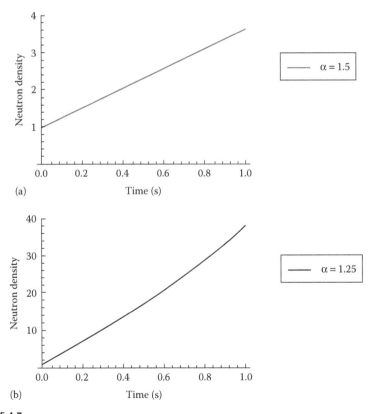

(a)

(b)

FIGURE 4.7
Neutron density for step reactivity (a) $\rho = 0.007\$$ with $\alpha = 1.5$ and (b) $\rho = 0.007\$$ with $\alpha = 1.25$.

(Continued)

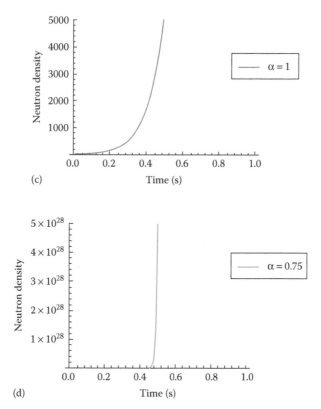

FIGURE 4.7 (Continued)
Neutron density for step reactivity (c) $\rho = 0.007\$$ with $\alpha = 1$ and (d) $\rho = 0.007\$$ with $\alpha = 0.75$.

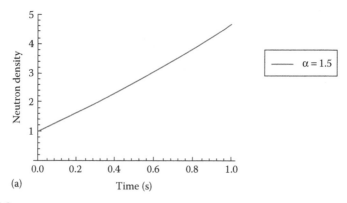

FIGURE 4.8
Neutron density for step reactivity (a) $\rho = 0.008\$$ with $\alpha = 1.5$.   *(Continued)*

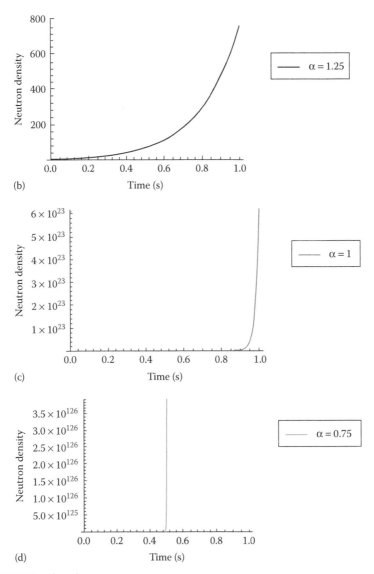

FIGURE 4.8 (Continued)
Neutron density for step reactivity (b) $\rho = 0.008\$$ with $\alpha = 1.25$, (c) $\rho = 0.008\$$ with $\alpha = 1$, and (d) $\rho = 0.008\$$ with $\alpha = 0.75$.

4.7.2 Results Obtained for Ramp Reactivity

Now we consider a ramp reactivity of $0.01\$/s$ and a neutron source free $q(t) = 0$ equilibrium system where the following parameters are used [53,64]:

$$\lambda_i = [0.0127, 0.0317, 0.115, 0.311, 1.4, 3.87]\,(\mathrm{s}^{-1})$$

$$\beta_i = [0.000266, 0.001491, 0.001316, 0.002849, 0.000896, 0.000182]$$

$$l = 0.00002\,\text{s}, \beta = 0.007$$

with the same initial condition as given in Equation 4.31.

The ramp reactivity function is $\rho(t) = 0.1\beta t$.

The numerical results for fractional neutron point kinetic equations with ramp reactivity (time-dependent function) are illustrated in Figures 4.9 through 4.11 by using different α. The comparison results between different α are shown in Table 4.10.

Hence, the numerical approximation results for neutron density obtained from the MDTM for ramp reactivity are in good agreement with the results obtained by McMohan and Pierson [64] with the help of Taylor series solution at classical order $\alpha = 1$. But using ramp reactivity there are no previous results existing in the open literature for fractional order point kinetic equations.

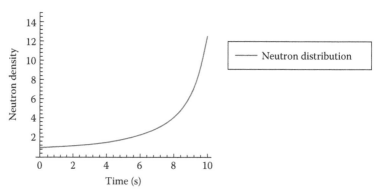

FIGURE 4.9
Neutron density with ramp reactivity with $\alpha = 1.5$ calculated using the MDTM.

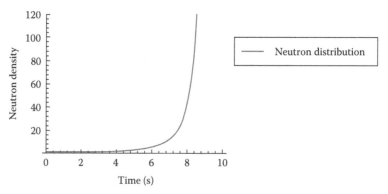

FIGURE 4.10
Neutron density with ramp reactivity with $\alpha = 1$ calculated using the MDTM.

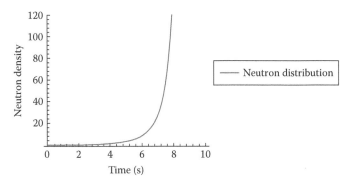

FIGURE 4.11
Neutron density with ramp reactivity with $\alpha = 0.75$ calculated using the MDTM.

TABLE 4.10

Results Obtained at Ramp Reactivity for Neutron Density Using the MDTM

	Neutron Density $n(t)$			McMohan and Pierson
Time (s)	$\alpha = 1.5$	$\alpha = 1$	$\alpha = 0.75$	(Classical Integer Order) [64]
$t = 2$	1.25661	1.3384	1.47755	1.3382
$t = 4$	1.72456	2.2287	3.40974	2.2285
$t = 6$	2.75227	5.5806	19.46960	5.5823
$t = 8$	6.15687	42.545	1336	42.7890
$t = 9$	12.61700	471.780	118901	487.520

4.7.3 Results Obtained for Sinusoidal Reactivity

Next we consider the last case involving sinusoidal reactivity [53,64]. In this case, we consider the following kinetic parameters:

$$\lambda_i = [0.0124, 0.0305, 0.111, 0.301, 1.14, 3.01]\,(\text{s}^{-1})$$

$$\beta_i = [0.000215, 0.001424, 0.001274, 0.002568, 0.000748, 0.000273]$$

$$l = 0.0005\,\text{s},\ \beta = 0.006502$$

The system is neutron source free $q(t) = 0$ with the same initial condition as given in Equation 4.31. The sinusoidal reactivity (time-dependent) function is $\rho(t) = \beta \sin(\pi t/T)$, where T is the half-life period ($T = 5$ s). The numerical results for fractional neutron point kinetic equations with sinusoidal reactivity (time-dependent function) are shown in Table 4.11 and also illustrated in Figures 4.12 through 4.14.

Thus the numerical approximation results for neutron density obtained from the MDTM for sinusoidal reactivity are in good agreement with the

TABLE 4.11

Results Obtained at Sinusoidal Reactivity for Neutron Density Using the MDTM

Time (s)	Neutron Density $n(t)$			McMohan and Pierson (Classical Integer Order) [64]
	$\alpha = 1.5$	$\alpha = 1$	$\alpha = 0.75$	
$t = 2$	8.42634	11.3325	12.9717	11.3820
$t = 4$	39.16030	90.0440	124.0650	92.2761
$t = 6$	6.69124	15.5705	21.5457	16.0317
$t = 8$	3.81628	8.4531	11.5648	8.6362
$t = 10$	5.97965	12.9915	17.7996	13.1987

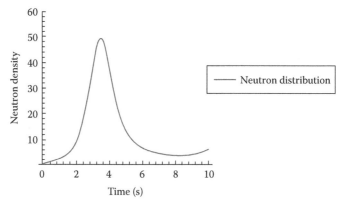

FIGURE 4.12

Neutron density with sinusoidal reactivity with $\alpha = 1.5$ obtained by the MDTM.

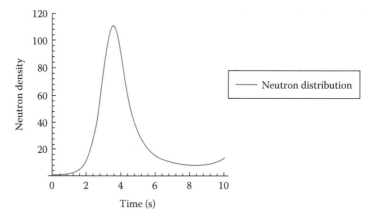

FIGURE 4.13

Neutron density with sinusoidal reactivity with $\alpha = 1$ obtained by the MDTM.

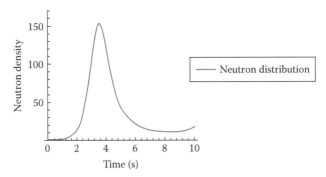

FIGURE 4.14
Neutron density with sinusoidal reactivity with $\alpha = 0.75$ obtained by the MDTM.

results obtained by McMohan and Pierson [64] with the help of Taylor series solution at classical order $\alpha = 1$. But using ramp reactivity there are no previous results existing in the open literature for fractional order point kinetic equations.

In this chapter, both classical and fractional order neutron point kinetic equations with arbitrary order α have been solved using the MDTM [66,67]. The present method is easier and more efficient to provide the numerical solution for classical as well as fractional neutron point kinetic equations. This method is a powerful solver for the classical and fractional neutron point kinetic equation. The present method is quite easy to apply for obtaining approximate numerical solutions for time-varying reactivities. Moreover, the accuracy can be further improved when the step size of each subinterval becomes smaller.

4.8 Conclusion

In this chapter, the classical and fractional order neutron point kinetic equations have been successfully solved by using the MDTM. From the obtained numerical results, it can be concluded that the MDTM is conveniently applicable to neutron point kinetic equation. Moreover, it also shows that the present method is reliable and promising when compared with other existing methods. The MDTM is clearly an effective and simple method for solving the classical and fractional order neutron point kinetic equation. It is extremely easy to apply the method. The method is more accurate to solve the problems with various types of reactivities. However, in this present method, the more accurate results can be obtained by taking more terms in the series and smaller time step size. Hence, the computational error is less.

5

Classical and Fractional Order Stochastic Neutron Point Kinetic Model

5.1 Introduction

Point kinetic equations represent the most important models in the field of nuclear science and engineering. The modeling of these equations characterizes the time-dependent behavior of a nuclear reactor [24,32,53,60]. Noise in reactors can be described by conventional point reactor kinetic equations (PRKEs) with fluctuation introduced in some of the parameters. Such equations may be referred to as stochastic PRKE. Power reactor noise analysis may be viewed as the study of a reactor's response to a stochastic reactivity or source input. The difficulty of solving stochastic PRKEs arises from the fact that they are nonlinear. The stochastic behavior of a point reactor is modeled with a system of Ito stochastic differential equations (SDEs).

It is well known that the reactions in a nuclear system are not fully describable by deterministic laws. This fact, at the most fundamental level, is due to the laws of quantum mechanics, which only give probabilities of various interactions for a neutron, which are manifest in the interaction cross sections of atoms with neutrons. There are various situations in which this probabilistic behavior could be readily observed for a nuclear system, for example, in the start-up of the reactors, in zero power reactors, and in most laboratory source-detector configurations. There has been an extensive research effort to model this stochastic behavior. Measuring higher-order moments requires more data from the system for a given accuracy. Actually, most of the times in practice, one only measures the first- and second-order moments in a system, that is, the mean and variance.

The standard deterministic point kinetic model has been the subject of countless studies and applications to understand the neutron dynamics and its effects, such as developed of different methods for their solution [49,50,68–71]. The reactivity function and neutron source term are the parametric quantities of this vital system. The dynamical process explained by the point kinetic equations is stochastic in nature. The neutron density and delayed neutron precursor concentrations differ randomly with respect to time. At the levels of high power, the random behavior is imperceptible. But at low-power levels, such as at the beginning, random fluctuation in the neutron density and neutron precursor concentrations can be crucial.

The numerical solutions for neutron population density and sum of precursors concentration population density have been solved with the stochastic piecewise constant approximation (PCA) method and Monte Carlo computations by using different step reactivity functions [60]. The derivation and the solution for stochastic neutron point kinetic equations have been elaborately described in the work of Saha Ray [71] by considering the same parameters and different step reactivity with Euler–Maruyama method and strong order 1.5 Taylor method. It can be observed that numerical methods like the Euler–Maruyama method and strong order 1.5 Taylor method are reliable like the stochastic PCA method and Monte Carlo computations. Here, the Euler–Maruyama method and strong order 1.5 Taylor method have been applied efficiently and conveniently for the solution of stochastic point kinetic equation with sinusoidal reactivity. The resulting systems of SDEs are solved over each time step size in the partition. In the present investigation, the main attractive advantage, of these computational numerical methods, is their simplicity, efficiency, and applicability.

In this research work, the numerical solution of fractional stochastic neutron point kinetic (FSNPK) equation has been obtained very efficiently and elegantly. In this work, for the first time ever, the random behavior of neutron density and neutron precursor concentrations have been analyzed in fractional order. Here, a numerical procedure has been used for efficiently calculating the solution for the FSNPK equation in the dynamical system of a nuclear reactor. The explicit finite difference method has been applied to solve the FSNPK equation with the Grünwald–Letnikov (GL) definition [44,72]. The FSNPK model has been analyzed for the dynamic behavior of the neutron.

5.2 Evolution of Stochastic Neutron Point Kinetic Model

The stochastic neutron point kinetic model is the most vital part of nuclear reactor dynamics; it is used to derive point kinetic equations in order to separate the processes of generation and extinction of neutrons in a reactor. It will help us to form a stochastic model. The deterministic time-dependent equations satisfied by the neutron density and the delayed neutron precursors are as follows [24]:

$$\frac{\partial N}{\partial t} = Dv\nabla^2 N - (\Sigma_a - \Sigma_f)vN + \left[(1-\beta)k_\infty\Sigma_a - \Sigma_f\right]vN + \sum_i \lambda_i C_i + S_0 \quad (5.1)$$

$$\frac{\partial C_i}{\partial t} = \beta_i k_\infty \Sigma_a vN - \lambda_i C_i, \quad i = 1, 2, \ldots, m \quad (5.2)$$

where:

$N(r,t)$ is the neutron density at a point r at time t

The coefficients D, v, Σ_a, and Σ_f are, respectively, diffusion constants, the neutron speed, the macroscopic neutron absorption, and the fission cross sections

$\Sigma_a - \Sigma_f$ is the capture cross section. If $\beta = \sum_{i=1}^{m} \beta_i$ be the delayed neutron fraction, the prompt neutron contribution to the source is $\left[(1-\beta)k_\infty \Sigma_a - \Sigma_f \right] vN$ and the prompt neutron fraction is $(1 - \beta)$

k_∞ is the number of neutrons produced per neutrons absorbed (also called infinite medium reproduction factor)

$\sum_{i=1}^{m} \lambda_i C_i$ is the rate of transformations from neutron precursors to the neutron population, where the delayed constant is λ_i and $C_i(r,t)$ is the density of the ith type of precursor for $i = 1,2,\cdots,m$

$S_0(r,t)$ is the source of neutrons extraneous to the fission process

In the present analysis, captures (or leakages) of neutrons are considered as deaths. The fission process is considered as pure birth process where $v(1 - \beta) - 1$ neutrons are born in each fission along with precursor fraction $v\beta$.

Let us assume that $N = f(r)n(t)$ and $C_i = g_i(r)c_i(t)$ are separable in time and space where $n(t)$ and $c_i(t)$ are the total number of neutrons and precursors of the ith type at time t, respectively.

Using these, Hetrick [24] and Hayes and Allen [60] derived the deterministic point kinetic equation as

$$\frac{dn}{dt} = -\left[\frac{-\rho + 1 - \alpha}{l} \right] n + \left[\frac{1 - \alpha - \beta}{l} \right] n + \sum_{i=1}^{m} \lambda_i c_i + q$$

$$\frac{dc_i}{dt} = \frac{\beta_i}{l} n - \lambda_i c_i, \quad i = 1,2,\cdots,m$$

(5.3)

where:

$q(t) = [S_0(r,t)/f(r)]$

ρ is reactivity

$l = 1/k_\infty v \Sigma_a$ is the neutron generation time

α is defined as $\alpha = \Sigma_f / \Sigma_a k_\infty \approx 1/v$

v is the average number of neutrons per fission

Here, $n(t)$ is the population size of neutrons and $c_i(t)$ is the population size of the ith neutron precursor. The neutron reactions can be separated into three terms as follows:

$$\frac{dn}{dt} = -\underbrace{\left[\frac{-\rho + 1 - \alpha}{l} \right] n}_{\text{Deaths}} + \underbrace{\left[\frac{1 - \alpha - \beta}{l} \right] n}_{\text{Births}} + \underbrace{\sum_{i=1}^{m} \lambda_i c_i}_{\text{Transformations}} + q$$

$$\frac{dc_i}{dt} = \frac{\beta_i}{l} n - \lambda_i c_i, \quad i = 1, 2, \cdots, m$$

The neutron birth rate due to fission is $b = (1 - \alpha - \beta)/l[-1 + (1 - \beta)v]$, where the denominator has the term $[-1 + (1 - \beta)v]$, which represents the number of neutrons (new born) produced in each fission process. The neutron death rate due to captures or leakage is $d = (-\rho + 1 - \alpha)/l$. The transformation rate $\lambda_i c_i$ represents the rate at which the ith precursor is transformed into neutrons and q represents the rate at which the source neutrons are produced.

To derive the stochastic dynamical system, we consider for simplicity only one precursor that is, $\beta = \beta_1$, where β is the total delayed neutron fraction for one precursor.

The point kinetic equations for one precursor are as follows:

$$\frac{dn}{dt} = \left[\frac{-\rho + 1 - \alpha}{l}\right] n + \left[\frac{1 - \alpha - \beta}{l}\right] n + \lambda_1 c_1 + q$$

$$\frac{dc_1}{dt} = \frac{\beta_1}{l} n - \lambda_1 c_1$$

Now, we consider the small duration of time interval Δt where probability of more than one occurred event is small. There are four different possibilities for an event at this small time Δt. Let $[\Delta n, \Delta c_1]^T$ be the change of n and c_1 in time Δt where the changes are assumed approximately normally distributed. The four possibilities for $[\Delta n, \Delta c_1]^T$ are as follows:

$$E_1 = \begin{bmatrix} \Delta n \\ \Delta c_1 \end{bmatrix}_1 \equiv \begin{bmatrix} -1 \\ 0 \end{bmatrix}$$

$$E_2 = \begin{bmatrix} \Delta n \\ \Delta c_1 \end{bmatrix}_2 \equiv \begin{bmatrix} -1 + (1 - \beta)v \\ \beta_1 v \end{bmatrix}$$

$$E_3 = \begin{bmatrix} \Delta n \\ \Delta c_1 \end{bmatrix}_3 \equiv \begin{bmatrix} 1 \\ -1 \end{bmatrix}$$

$$E_4 = \begin{bmatrix} \Delta n \\ \Delta c_1 \end{bmatrix}_4 \equiv \begin{bmatrix} 1 \\ 0 \end{bmatrix}$$

where:
E_1 is the first event denoting death
E_2 is the second event representing birth of $[-1 + (1 - \beta)v]$ neutrons
$\beta_1 v$ delayed neutron precursors produced in fission process

E_3 is the third event representing a transformation of a delayed neutron precursor to a neutron

E_4 is the last event representing a neutron source

The respective probabilities of these events are as follows:

$$P(E_1) = n\Delta t d$$

$$P(E_2) = n\Delta t b = \frac{1}{vl} n\Delta t, \quad \text{since } b = \frac{1-\alpha-\beta}{l\left[-1+(1-\beta)v\right]} \text{ and } \alpha = \frac{\Sigma_f}{\Sigma_a k_\infty} \approx \frac{1}{v}$$

$$P(E_3) = c_1 \Delta t \lambda_1$$

$$P(E_4) = q\Delta t$$

In this present analysis, it is assumed that the extraneous source randomly produces neutrons that follow the Poisson process with intensity q.

According to our earlier assumption, the changes in neutron population and precursor concentration are approximately normally distributed with

mean $E\left(\begin{bmatrix} \Delta n \\ \Delta c_1 \end{bmatrix}\right)$ and variance $\text{Var}\left(\begin{bmatrix} \Delta n \\ \Delta c_1 \end{bmatrix}\right)$.

Here, the mean change in the small interval of time Δt is

$$E\left(\begin{bmatrix} \Delta n \\ \Delta c_1 \end{bmatrix}\right) = \sum_{k=1}^{4} P_k \begin{bmatrix} \Delta n \\ \Delta c_1 \end{bmatrix}_k = \begin{bmatrix} \frac{\rho-\beta}{l} n + \lambda_1 c_1 + q \\ \frac{\beta_1}{l} n - \lambda_1 c_1 \end{bmatrix} \Delta t$$

and the variance of change in small time Δt is

$$\text{Var}\left(\begin{bmatrix} \Delta n \\ \Delta c_1 \end{bmatrix}\right) = E\left(\begin{bmatrix} \Delta n \\ \Delta c_1 \end{bmatrix} \begin{bmatrix} \Delta n & \Delta c_1 \end{bmatrix}\right) - \left(E\left\{\begin{bmatrix} \Delta n \\ \Delta c_1 \end{bmatrix}\right\}\right)^2$$

$$= \sum_{k=1}^{4} P_k \begin{bmatrix} \Delta n \\ \Delta c_1 \end{bmatrix}_k \begin{bmatrix} \Delta n & \Delta c_1 \end{bmatrix}_k = \hat{B}\Delta t$$

where:

$$\hat{B} = \begin{bmatrix} \gamma n + \lambda_1 c_1 + q & \frac{\beta_1}{l}\left(-1+(1-\beta)v\right)n - \lambda_1 c_1 \\ \frac{\beta_1}{l}\left(-1+(1-\beta)v\right)n - \lambda_1 c_1 & \frac{\beta_1^2 v}{l} n + \lambda_1 c_1 \end{bmatrix}$$

where:

$$\gamma = \frac{-1-\rho+2\beta+(1-\beta)^2 v}{l}$$

Now, by central limit theorem, the random variate

$$\left\{ \begin{bmatrix} \Delta n \\ \Delta c_1 \end{bmatrix} - E\left(\begin{bmatrix} \Delta n \\ \Delta c_1 \end{bmatrix} \right) \right\} \Big/ \sqrt{\mathrm{Var}\left(\begin{bmatrix} \Delta n \\ \Delta c_1 \end{bmatrix} \right)}$$

follows the standard normal distribution.
The above result implies the following:

$$\begin{bmatrix} \Delta n \\ \Delta c_1 \end{bmatrix} = E\left(\begin{bmatrix} \Delta n \\ \Delta c_1 \end{bmatrix} \right) + \sqrt{\mathrm{Var}\left(\begin{bmatrix} \Delta n \\ \Delta c_1 \end{bmatrix} \right)} + \begin{bmatrix} \eta_1 \\ \eta_2 \end{bmatrix}, \quad \text{where } \eta_1, \eta_2 \sim N(0,1) \qquad (5.4)$$

Thus, we have

$$\begin{bmatrix} n(t+\Delta t) \\ c_1(t+\Delta t) \end{bmatrix} = \begin{bmatrix} n(t) \\ c_1(t) \end{bmatrix} + \begin{bmatrix} \dfrac{\rho-\beta}{l}n + \lambda_1 c_1 \\[2mm] \dfrac{\beta_1}{l}n + \lambda_1 c_1 \end{bmatrix} \Delta t + \begin{bmatrix} q \\ 0 \end{bmatrix} \Delta t + \hat{B}^{1/2}\sqrt{\Delta t}\begin{bmatrix} \eta_1 \\ \eta_2 \end{bmatrix} \qquad (5.5)$$

where:
$\hat{B}^{1/2}$ is the square root of the matrix $\hat{B}$

Dividing both sides of Equation 5.5 by Δt and then taking limit $\Delta t \to 0$, we achieve the following Itô SDE system

$$\frac{d}{dt}\begin{bmatrix} n \\ c_1 \end{bmatrix} = \hat{A}\begin{bmatrix} n \\ c_1 \end{bmatrix} + \begin{bmatrix} q \\ 0 \end{bmatrix} + \hat{B}^{1/2}\frac{d\overrightarrow{W}}{dt} \qquad (5.6)$$

where:

$$\hat{A} = \begin{bmatrix} \dfrac{\rho-\beta}{l} & \lambda_1 \\[3mm] \dfrac{\beta_1}{l} & -\lambda_1 \end{bmatrix}$$

$$\hat{B} = \begin{bmatrix} \gamma n + \lambda_1 c_1 + q & \dfrac{\beta_1}{l}\left[-1+(1-\beta)v\right]n - \lambda_1 c_1 \\[4mm] \dfrac{\beta_1}{l}\left[-1+(1-\beta)v\right]n - \lambda_1 c_1 & \dfrac{\beta_1^2}{l}n + \lambda_1 c_1 \end{bmatrix}$$

and

$$\overline{W}(t) = \begin{bmatrix} W_1(t) \\ W_2(t) \end{bmatrix} = \lim_{\Delta t \to 0} \frac{1}{\sqrt{\Delta t}} \begin{bmatrix} \eta_1 \\ \eta_2 \end{bmatrix}, \quad \text{for } \eta_1, \eta_2 \sim N(0,1)$$

where:

$W_1(t)$ and $W_2(t)$ are Wiener process

Equation 5.6 represents the stochastic point kinetic equations for one precursor.

Now generalizing the above argument to m precursors, we can obtain the following Itô SDE system for m precursors:

$$\frac{d}{dt} \begin{bmatrix} n \\ c_1 \\ c_2 \\ \vdots \\ c_m \end{bmatrix} = \hat{A} \begin{bmatrix} n \\ c_1 \\ c_2 \\ \vdots \\ c_m \end{bmatrix} + \begin{bmatrix} q \\ 0 \\ 0 \\ \vdots \\ 0 \end{bmatrix} + \hat{B}^{1/2} \frac{d\overline{W}}{dt} \tag{5.7}$$

In the above equation, $\hat{A}$ and $\hat{B}$ are as follows:

$$\hat{A} = \begin{bmatrix} \dfrac{\rho - \beta}{l} & \lambda_1 & \lambda_2 & \cdots & \lambda_m \\ \dfrac{\beta_1}{l} & -\lambda_1 & 0 & \cdots & 0 \\ \dfrac{\beta_2}{l} & 0 & -\lambda_2 & \ddots & \vdots \\ \vdots & \vdots & \ddots & \ddots & 0 \\ \dfrac{\beta_m}{l} & 0 & \cdots & 0 & -\lambda_m \end{bmatrix} \tag{5.8}$$

$$\hat{B} = \begin{bmatrix} \zeta & a_1 & a_2 & \cdots & a_m \\ a_1 & r_1 & b_{2,3} & \cdots & b_{2,m+1} \\ a_2 & b_{3,2} & r_2 & \ddots & \vdots \\ \vdots & \vdots & \ddots & \ddots & b_{m,m+1} \\ a_m & b_{m+1,2} & \cdots & b_{m+1,m} & r_m \end{bmatrix}$$

where:

$$\zeta = \gamma n + \sum_{j=1}^{m} \lambda_j c_j + q$$

$$\gamma = \frac{-1-\rho+2\beta+(1-\beta)^2 v}{l}$$

$$a_j = \frac{\beta_j}{l}\left[-1+(1-\beta)v\right]n - \lambda_j c_j$$

$$b_{i,j} = \frac{\beta_{i-1}\beta_{j-1}v}{l}n$$

and

$$r_i = \frac{\beta_i^2 v}{l}n + \lambda_i c_i$$

Equation 5.7 represents the generalization of the standard point kinetic model; since for $\hat{B} = 0$, it reduces to the standard deterministic point kinetic model [53].

5.3 Classical Order Stochastic Neutron Point Kinetic Model

A point reactor is a reactor in which the spatial effects have been eliminated. This is obviously possible if the reactors length is infinite in all spatial dimensions. Study of a point reactor, that is, studying the properties of Equation 5.9, is desirable in the sense that it captures some of the most essential features of the reactor dynamics without involving into the complexities of integro-differential equations, that is, the transport equation, or partial differential equations, that is, the diffusion equation.

In order to separate the birth and death process of neutron population, Hetrick [24] and Hayes and Allen [60] derived the deterministic point kinetic equation as

$$\frac{dn}{dt} = -\left[\frac{-\rho+1-\alpha}{l}\right]n + \left[\frac{1-\alpha-\beta}{l}\right]n + \sum_{i=1}^{m}\lambda_i c_i + q$$

(5.9)

$$\frac{dc_i}{dt} = \frac{\beta_i}{l}n - \lambda_i c_i, \quad i = 1,2,\cdots,m$$

Here, $n(t)$ is the population size of neutrons and $c_i(t)$ is the population size of the ith neutron precursor.

Now, the stochastic point kinetic equations for m-delayed groups is defined as [71]

$$\frac{d\vec{x}}{dt} = A\vec{x} + B(t)\vec{x} + \vec{F}(t) + \hat{B}^{1/2}\frac{d\overrightarrow{W}}{dt} \tag{5.10}$$

with initial condition $\vec{x}(0) = \vec{x}_0$. Here

$$\vec{x} = \begin{bmatrix} n \\ c_1 \\ c_2 \\ \vdots \\ c_m \end{bmatrix} \tag{5.11}$$

and $\hat{B}$ is given in Equation 5.6.

A is $(m+1)\times(m+1)$ matrix given by

$$A = \begin{bmatrix} \dfrac{-\beta}{l} & \lambda_1 & \lambda_2 & \cdots & \lambda_m \\ \dfrac{\beta_1}{l} & -\lambda_1 & 0 & \cdots & 0 \\ \dfrac{\beta_2}{l} & 0 & -\lambda_2 & \ddots & \vdots \\ \vdots & \vdots & \ddots & \ddots & 0 \\ \dfrac{\beta_m}{l} & 0 & \cdots & 0 & -\lambda_m \end{bmatrix} \tag{5.12}$$

B is $(m+1)\times(m+1)$ matrix given by

$$B = \begin{bmatrix} \dfrac{\rho(t)}{l} & 0 & 0 & \cdots & 0 \\ 0 & 0 & 0 & \cdots & 0 \\ 0 & 0 & 0 & \ddots & \vdots \\ \vdots & \vdots & \ddots & \ddots & 0 \\ 0 & 0 & \cdots & 0 & 0 \end{bmatrix} \tag{5.13}$$

and $\vec{F}(t)$ is given as

$$\vec{F}(t) = \begin{bmatrix} q(t) \\ 0 \\ 0 \\ \vdots \\ 0 \end{bmatrix} \tag{5.14}$$

It can be noticed that $\hat{A} = A + B(t)$ where A is a constant matrix.

5.4 Numerical Solution of the Classical Stochastic Neutron Point Kinetic Equation

5.4.1 Euler–Maruyama Method for the Solution of Stochastic Point Kinetic Model

The Euler–Maruyama approximation is the simplest time discrete approximations of an Itô process. Let $\{Y_\tau\}$ be an Itô process on $\tau \in \left[t_0, T\right]$ satisfying the SDE

$$\begin{cases} dY_\tau = a(\tau,\, Y_\tau)d\tau + b(\tau,\, Y_\tau)dW_\tau \\ Y_{t0} = Y_0 \end{cases} \tag{5.15}$$

For a given time discretization

$$t_0 = \tau_0 < \tau_1 < \cdots < \tau_n = T \tag{5.16}$$

an Euler approximation is a continuous time stochastic process $\left\{X(\tau),\, t_0 \le \tau \le T\right\}$ satisfying the iterative scheme

$$X_{n+1} = X_n + a(\tau_n,\, X_n)\Delta\tau_{n+1} + b(\tau_n,\, X_n)\Delta W_{n+1}, \quad \text{for } n = 0,1,2,\ldots,N-1 \tag{5.17}$$

where:

$$X_n = X(\tau_n)$$

$$\Delta\tau_{n+1} = \tau_{n+1} - \tau_n$$

$$\Delta W_{n+1} = W(\tau_{n+1}) - W(\tau_n)$$

with initial value $X_0 = X(\tau_0)$. Here, each random number ΔW_n is computed as $\Delta W_n = \eta_n \sqrt{\Delta\tau_n}$ where η_n is chosen from standard normal distribution $N(0,1)$.

We have considered the equidistant discretized times $\tau_n = \tau_0 + n\Delta$ with $\Delta = \Delta_n = (T - \tau_0)/N$ for some integer N large enough so that $\Delta \in (0,1)$.

The Euler–Maruyama method is also known as strong order 0.5 Itô–Taylor approximation. By applying the Euler–Maruyama method to Equation 5.10 in view of Equation (5.15), we obtain the scheme as

$$\vec{x}_{i+1} = \vec{x}_i + (A + B_i)\vec{x}_i h + \vec{F}(t_i)h + \hat{B}^{1/2}\sqrt{h}\,\vec{\eta}_i \tag{5.18}$$

where:
$d\vec{W}_i = \vec{W}_i - \vec{W}_{i-1} = \sqrt{h}\,\vec{\eta}_i$
$h = t_{i+1} - t_i$
$\vec{\eta}_i$ is a vector whose components are random numbers chosen from $N(0,1)$

5.4.2 Strong Order 1.5 Taylor Method for the Solution of Stochastic Point Kinetic Model

Here we consider Taylor approximation having strong order $\alpha = 1.5$. The strong order 1.5 Taylor scheme can be obtained by adding more terms from

Itô–Taylor expansion to the Milstein scheme [19,21]. The strong order 1.5 Itô–Taylor scheme is

$$Y_{n+1} = Y_n + a\Delta_n + b\Delta W_n + \frac{1}{2}bb_x\left(\Delta W_n^2 - \Delta_n\right) + a_x b\Delta Z_n + \frac{1}{2}\left(aa_x + \frac{1}{2}b^2 a_{xx}\right)\Delta_n^2$$

$$\tag{5.19}$$

$$+ \left(ab_x + \frac{1}{2}b^2 b_{xx}\right)(\Delta W_n \Delta_n - \Delta Z_n) + \frac{1}{2}b\left(bb_{xx} + b_x^2\right)\left(\frac{1}{3}\Delta W_n^2 - \Delta_n\right)\Delta W_n$$

for $n = 0,1,2,\ldots,N-1$
   with initial value

$$Y_0 = Y(\tau_0) \text{ and } \Delta_n = \Delta\tau_n$$

Here, partial derivatives are denoted by subscripts and the random variable ΔZ_n is normally distributed with mean $E(\Delta Z_n) = 0$ and variance $E\left(\Delta Z_n^2\right) = (1/3)\Delta\tau_n^3$ and correlated with ΔW_n by covariance $E\left(\Delta Z_n \Delta W_n\right) = (1/2)\Delta\tau_n^2$.
   We can generate ΔZ_n as

$$\Delta Z_n = \frac{1}{2}\Delta\tau_n\left(\Delta W_n + \frac{\Delta V_n}{\sqrt{3}}\right) \tag{5.20}$$

where ΔV_n is chosen independently from $\sqrt{\Delta\tau_n}\, N(0,1)$. Here, the approximation $Y_n = Y(\tau_n)$ is the continuous time stochastic process $\{Y(\tau), t_0 \le \tau \le T\}$, the time step size $\Delta\tau_n = \tau_n - \tau_{n-1}$ and $\Delta W_n = W(\tau_n) - W(\tau_{n-1})$.
   By applying the strong order 1.5 Taylor approximation methods to Equation 5.10 in view of Equation 5.19 yielding

$$\vec{x}_{i+1} = \vec{x}_i + \left[(A + B_i)\vec{x}_i + \vec{F}_i\right]h + \hat{B}^{1/2}\sqrt{h}\,\vec{\eta}_i + (A + B_i)\hat{B}^{1/2}\Delta Z_i$$

$$\tag{5.21}$$

$$+ \frac{1}{2}\left[(A + B_i)\vec{x}_i + \vec{F}_i\right](A + B_i)h^2$$

where:

$$\Delta Z_i = \frac{1}{2}h\left(\Delta W_i + \frac{\Delta V_i}{\sqrt{3}}\right)$$

$$\Delta V_i = \sqrt{h}\, N(0,1)$$

5.5 Numerical Results and Discussions for the Solution of Stochastic Point Kinetic Model

In the present analysis, we consider the first example of a nuclear reactor problem with the following parameters $\lambda_1 = 0.1$, $\beta_1 = 0.05 = \beta$, $v = 2.5$, neutron source $q = 200$, $l = 2/3$, and $\rho(t) = -1/3$ for $t \geq 0$. The initial condition is $\bar{x}(0) = \begin{bmatrix} 400 & 300 \end{bmatrix}^T$. We observe through 5000 trails the good agreement between two methods with other available methods for 40 time intervals at time $t = 2$ s. The mean and standard deviation of $n(2)$ and $c_1(2)$ are presented in Table 5.1.

In the second example, we assume the initial condition as

$$\bar{x}(0) = 100 \begin{bmatrix} 1 \\ \dfrac{\beta_1}{\lambda_1 l} \\ \dfrac{\beta_2}{\lambda_2 l} \\ \vdots \\ \dfrac{\beta_m}{\lambda_m l} \end{bmatrix}$$

The following parameters are used in the works of Hetrick [24] and Kinard and Allen [53]: $\beta = 0.007$, $v = 2.5$, $l = 0.00002$, $q = 0$, $\lambda_i = [0.0127, 0.0317, 0.115, 0.311, 1.4, 3.87]$, and $\beta_i = [0.000266, 0.001491, 0.001316, 0.002849, 0.000896, 0.000182]$ with $m = 6$ delayed groups. The computational results at $t = 0.1$ and 0.001 are given in Tables 5.2 and 5.3, respectively, for the Monte Carlo, stochastic PCA [60], Euler–Maruyama, and strong order 1.5 Taylor methods. It can be seen that there exists approximately close agreements between the three approaches in consideration with different step reactivities $\rho = 0.003$ and 0.007. The mean neutron density and two individual neutron samples

TABLE 5.1

Comparison of Numerical Computational Methods for One Precursor

	Monte Carlo	Stochastic PCA [60]	Euler–Maruyama Approximation	Strong Order 1.5 Taylor Approximation
$E[n(2)]$	400.03	395.32	412.23	412.10
$\sigma[n(2)]$	27.311	29.411	34.391	34.519
$E[c_1(2)]$	300.00	300.67	315.96	315.93
$\sigma[c_1(2)]$	7.8073	8.3564	8.2656	8.3158

TABLE 5.2

Comparison for Subcritical Step Reactivity $\rho = 0.003$

	Monte Carlo	Stochastic PCA [60]	Euler–Maruyama Approximation	Strong Order 1.5 Taylor Approximation
$E[n(0.1)]$	183.04	186.31	208.599	199.408
$\sigma[n(0.1)]$	168.79	164.16	255.954	168.547
$E[c_1(0.1)]$	4.478×10^5	4.491×10^5	4.498×10^5	4.497×10^5
$\sigma[c_1(0.1)]$	1495.7	1917.2	1233.38	1218.82

TABLE 5.3

Comparison for Critical Step Reactivity $\rho = 0.007$

	Monte Carlo	Stochastic PCA [60]	Euler–Maruyama Approximation	Strong Order 1.5 Taylor Approximation
$E[n(0.001)]$	135.67	134.55	139.568	139.569
$\sigma[n(0.001)]$	93.376	91.242	92.042	92.047
$E[c_1(0.001)]$	4.464×10^5	4.464×10^5	4.463×10^5	4.463×10^5
$\sigma[c_1(0.001)]$	16.226	19.444	6.071	18.337

are illustrated in Figure 5.1 and the mean precursor density and two precursor sample paths are illustrated in Figure 5.2. For these calculations, we used 5000 trials in both the Euler–Maruyama method and strong order 1.5 Taylor method.

Again, the one-group delayed neutron point kinetic model of a reactor has been considered. In the present analysis, reactivity has been considered as sinusoidal reactivity function $\rho(t) = \rho_0 \sin(\pi t/T)$ in a nuclear reactor [24,53] with $m = 1$ delayed group and the parameters are as follows: $\rho_0 = 0.005333$ (68 cents), $\lambda_1 = 0.077$ s^{-1}, $\beta_1 = 0.0079 = \beta$, neutron source $q(t) = 0$, $l = 10^{-3}$ s, a half-life period $T = 50$ s, and $t \geq 0$. The initial condition is $\vec{x}(0) = \begin{bmatrix} 1 & (\beta_1/\lambda_1 l) \end{bmatrix}^T$. We observe through a period of 800 s that there is a good agreement available between the two present methods, namely, the Euler–Maruyama method and strong order 1.5 Taylor method. The computational results are shown in Figures 5.3 and 5.4 for neutron population density with respect to time with different time step size at different trials. The neutron population density obtained by the Euler–Maruyama method and strong order 1.5 Taylor method using a sinusoidal reactivity for $t = 800$ s with step size $h = 0.001$ at 100 trials is illustrated in Figure 5.3a and 5.3b. Then, we reduce the number of trials to 30, and the neutron population density obtained by the Euler–Maruyama method and strong order 1.5 Taylor method using a sinusoidal reactivity for $t = 800$ s with step size $h = 0.001$ is illustrated in Figure 5.3c and 5.3d. Similarly in Figure 5.4a and 5.4b, the numerical computation of the neutron population density obtained by the Euler–Maruyama method and strong order 1.5 Taylor method for $t = 800$ s with step size $h = 0.01$ at 100 trials is plotted. We have considered two random

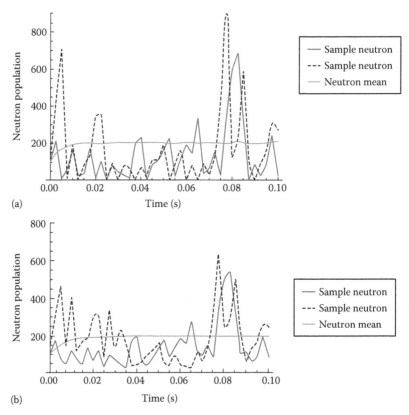

(a)

(b)

FIGURE 5.1
(a) Neutron density obtained by the Euler–Maruyama method using a subcritical step reactivity $\rho = 0.003$, (b) neutron density obtained by the strong order 1.5 Taylor method using a subcritical step reactivity $\rho = 0.003$.

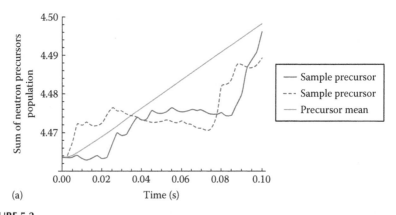

(a)

FIGURE 5.2
(a) Precursor density obtained by the Euler–Maruyama method using a subcritical step reactivity $\rho = 0.003$.   (*Continued*)

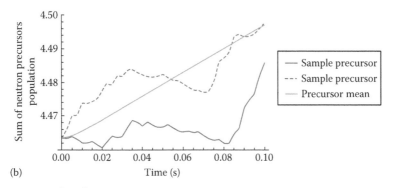

(b)

FIGURE 5.2 (Continued)
(b) Precursor density obtained by the strong order 1.5 Taylor method using a subcritical step reactivity $\rho = 0.003$.

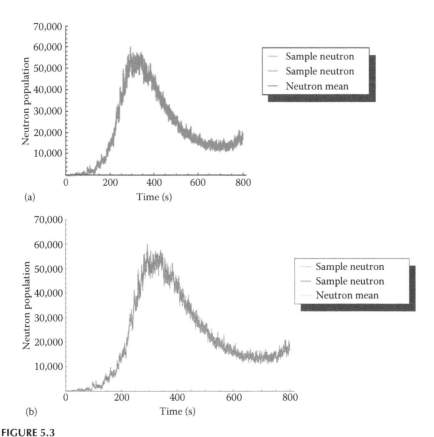

FIGURE 5.3
(a) Neutron population density obtained by the Euler–Maruyama method with step size $h = 0.001$ at 100 trials, (b) neutron density obtained by the strong order 1.5 Taylor method with step size $h = 0.001$ at 100 trials.   (*Continued*)

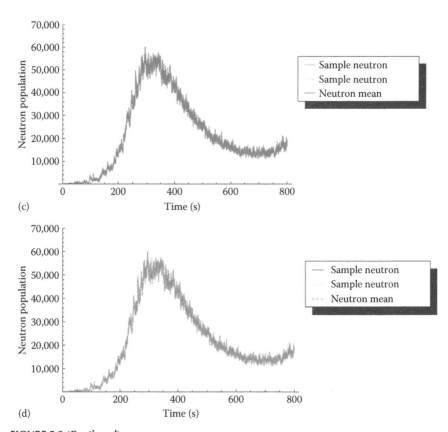

FIGURE 5.3 (Continued)
(c) Neutron population density obtained by the Euler–Maruyama method with step size $h = 0.001$ at 30 trials, (d) neutron density obtained by the strong order 1.5 Taylor method with step size $h = 0.001$ at 30 trials.

samples of neutrons at the 25th and 30th trials of the total sample of 5000 trial for the solution. In comparison with the deterministic point kinetic model with sinusoidal reactivity given in the work of Kinard and Allen [53], the present numerical methods are more efficient and accurate to give the solution of stochastic neutron point kinetic equation with sinusoidal reactivity for one precursor.

Next, it has been considered the effect of pulse reactivity function [24,32,64]

$$\rho(t) = \begin{cases} 4\beta(e^{-2t^2}), & t < 1 \\ 0, & \text{otherwise} \end{cases}$$

in a nuclear reactor with $m = 1$ neutron precursor and the parameters used are as follows: $\lambda_1 = 0.077 \text{ s}^{-1}$, $\beta_1 = 0.006502 = \beta$, neutron source $q = 0$, and $l = 10^{-3} \text{ s}$ [64]. Mathematically, we can define neutron mean

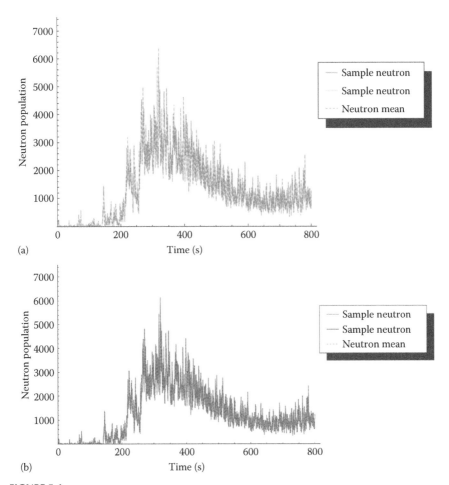

FIGURE 5.4
(a) Neutron population density obtained by the Euler–Maruyama method with step size $h = 0.01$ at 100 trials, (b) Neutron population density obtained by the strong order 1.5 Taylor method with step size $h = 0.01$ at 100 trials.

$$E[n(t)] = (1/N) \sum_{j=1}^{N} n_j(t)$$

and similarly mean of m-group precursor

$$E[c(t)] = (1/N) \sum_{j=1}^{N} \left[\sum_{k=1}^{m} c_{jk}(t) \right]$$

where N is the total number of trials. The initial condition is $\bar{x}(0) = \begin{bmatrix} 1 & (\beta_1/\lambda_1 l) \end{bmatrix}^T$.
We observe the behavior through a period of 1 s. The obtained numerical

TABLE 5.4

Comparison between Numerical Computational Methods for One Precursor

	Euler–Maruyama Approximation	Strong Order 1.5 Taylor Approximation
$E[n(0.001)]$	1.98979	1.99135
$E[c_1(0.001)]$	84.528	84.5285
$E[n(0.1)]$	522.98	525.137
$E[c_1(0.1)]$	212.141	212.78
$E[n(1)]$	1.42861×10^6	1.43966×10^6
$E[c_1(1)]$	4.40479×10^6	4.4397×10^6

approximation results for mean neutron population $E[n(t)]$ and mean of m-group precursor population $E[c(t)]$ are given in Table 5.4 for time 0.001, 0.1, and 1 s with step size $h = 0.0001$ at single trial. The mean of neutron population density and the mean of sum of precursor density obtained by the Euler–Maruyama method and strong order 1.5 Taylor method using a pulse reactivity for $t = 1$ s with step size $h = 0.01$ at 100 trials are illustrated in Figures 5.5 through 5.8.

The main advantage of the Euler–Maruyama method and strong order 1.5 Taylor method is first the PCA [53,60] over a partition is not required for reactivity and source functions. There is no need to obtain the eigenvalues and eigenvectors of point kinetic matrices for getting the solution of stochastic point kinetic equations. Use of inhour equation to calculate eigenvalues by finding roots of polynomial function was presented in the work of Hetrick [24]. But it is a very cumbersome and complicated procedure. In the present methods, such types of unwieldy computations are not required.

The limitation of the Euler–Maruyama method is that its order of convergence is $\alpha = 1/2$, which is slower in comparison with the strong order 1.5 Taylor method whose order of convergence is $\alpha = 3/2$.

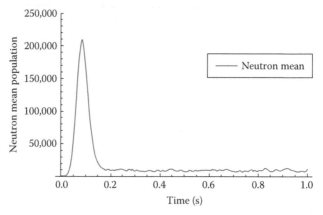

FIGURE 5.5
Neutron mean population density obtained by using the Euler–Maruyama method.

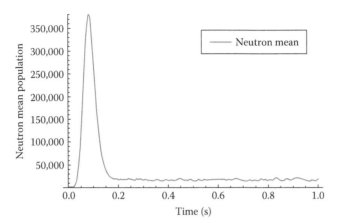

FIGURE 5.6
Neutron mean population density obtained by using the strong order 1.5 Taylor method.

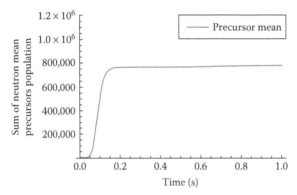

FIGURE 5.7
Sum of neutron mean precursor density obtained by using the Euler–Maruyama method.

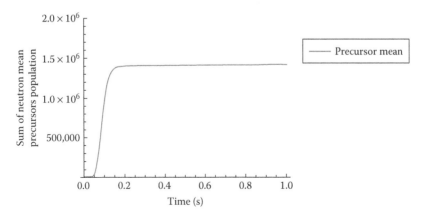

FIGURE 5.8
Sum of neutron mean precursor density obtained by using the strong order 1.5 Taylor method.

5.6 Application of Explicit Finite Difference Method for Solving Fractional Order Stochastic Neutron Point Kinetic Model

We consider the neutron point kinetic equations with GL fractional time derivative [49,50,68–70] and having behavior of Itô-type stochastic system [60,71] for m-group delayed neutron precursors, in the field of nuclear reactor dynamics, as follows:

$$\frac{d^\alpha \vec{x}}{dt^\alpha} = A\vec{x} + B(t)\vec{x} + \vec{F}(t) + \hat{B}^{1/2}\frac{d\vec{W}}{dt}, \quad 0 < \alpha \le 1 \tag{5.22}$$

subject to initial condition $\vec{x}(0) = \vec{x}_0$.

In the nuclear reactor, at the lower-power levels during the start-up of a nuclear reactor operation, the random fluctuation in the neutron population density and neutron precursor concentrations has been observed and it is significant. Random behavior of the neutron population and precursors concentrations played a vital role in the process of diffusion that occurred inside the reactor [60]. If $\alpha = 1$, the process is normal diffusion, and when $0 < \alpha < 1$, the process is anomalous diffusion.

To examine the behavior of anomalous diffusion, we have considered the stochastic neutron point kinetic equation in fractional order. The study of random nature and anomalous diffusion has not been investigated by any researchers for the FSNPK equations.

Here,

$$\vec{x}(t) = \begin{bmatrix} n(t) \\ c_1(t) \\ c_2(t) \\ \vdots \\ c_m(t) \end{bmatrix}$$

$$A = \begin{bmatrix} \dfrac{-\beta}{l} & \lambda_1 & \lambda_2 & \cdots & \lambda_m \\ \dfrac{\beta_1}{l} & -\lambda_1 & 0 & \cdots & 0 \\ \dfrac{\beta_2}{l} & 0 & -\lambda_2 & \ddots & \vdots \\ \vdots & \vdots & \ddots & \ddots & 0 \\ \dfrac{\beta_m}{l} & 0 & \cdots & 0 & -\lambda_m \end{bmatrix}_{(m+1)\times(m+1)} \tag{5.23}$$

$$B = \begin{bmatrix} \dfrac{\rho(t)}{l} & 0 & 0 & \cdots & 0 \\ 0 & 0 & 0 & \cdots & 0 \\ 0 & 0 & 0 & \ddots & \vdots \\ \vdots & \vdots & \ddots & \ddots & 0 \\ 0 & 0 & \cdots & 0 & 0 \end{bmatrix}_{(m+1) \times (m+1)}$$

(5.24)

and $\vec{F}(t)$ is given as

$$\vec{F}(t) = \begin{bmatrix} q(t) \\ 0 \\ 0 \\ \vdots \\ 0 \end{bmatrix}$$

(5.25)

Let us take the time step size h. Using the definition of GL fractional derivative, the numerical approximation of the Equation 5.22, in view of the research work [13,50,52], is

$$h^{-\alpha} \sum_{j=0}^{m} \omega_j^{(\alpha)} \vec{x}_{(m-j)} = A\vec{x}_{(m)} + B\vec{x}_{(m)} + \vec{F}_{(m)} + \hat{B}^{1/2} h^{-1/2} \vec{\eta}_{(m)}$$

(5.26)

The above equation leads to implicit numerical iteration scheme. However, in this present work, we propose an explicit numerical scheme that leads from the time layer t_{m-1} to t_m as follows:

$$\vec{x}_{(m)} = -\sum_{j=1}^{m} \omega_j^{(\alpha)} \vec{x}_{(m-j)} + h^{\alpha} \left[A\vec{x}_{(m-1)} + B\vec{x}_{(m-1)} + \vec{F}_{(m)} + \hat{B}^{1/2} h^{-1/2} \vec{\eta}_{(m)} \right]$$

(5.27)

where:
$\vec{x}_m = \vec{x}(t_m)$ is the mth approximation at time t_m
$t_m = mh, \ m = 0,1,2,\ldots$

$$\omega_j^{\alpha} = (-1)^j \binom{\alpha}{j}, \ j = 0,1,2,3,\cdots$$

Here, $d\vec{W} = \vec{W}_{m+1} - \vec{W}_m = \Delta \vec{W}_m = \sqrt{h} \vec{\eta}_m$ where $\vec{\eta}_m$ is chosen from $N(0,1)$ and $h = t_{m+1} - t_m$ with initial condition $\vec{x}_0 = \vec{x}(t_0)$.

5.7 Numerical Results and Discussions for the FSNPK Equations

In this section, we consider the first example of nuclear reactor problems [60], $\lambda_1 = 0.1\,\text{s}^{-1}$, $\beta_1 = 0.05 = \beta$, $v = 2.5$, neutron source $q = 200$, $l = 2/3\,\text{s}$, and $\rho(t) = -1/3$ for $t \geq 0$. The initial condition is $\bar{x}(0) = \begin{bmatrix} 400 & 300 \end{bmatrix}^T$. We applied explicit finite difference methods for 40 intervals at time $t = 2$ s. The mean and standard deviation of $n(2)$ and $c_1(2)$ are presented in Table 5.5 for different values of fractional order α. The numerical results for the time period of $t = 1$ s are given graphically with positive and negative reactivity in Figures 5.9 through 5.12. For negative reactivity at $\rho(t) = -1/3$, it can be observed from Figures 5.9 and 5.10, the neutron population decreases with small time variation, and for positive reactivity at $\rho(t) = 1/3$, the neutron population gradually increases with time in Figures 5.11 and 5.12.

Moreover, for negative reactivity $\rho(t) = -1/3$, Figures 5.9 and 5.10 exhibit the comparison of neutron population behavior in case of fractional order 1/2 and classical integer order, respectively. Similarly, for positive reactivity $\rho(t) = 1/3$, Figures 5.11 and 5.12 exhibit the comparison of neutron population behavior in case of fractional order 1/2 and classical integer order, respectively.

TABLE 5.5

Mean and Standard Deviation of Neutron and Precursor for Different Values of α

	$\alpha = 0.25$	$\alpha = 0.5$	$\alpha = 0.75$	$\alpha = 1$ (Classical)
$E[n(2)]$	162.236	201.703	267.868	412.23
$\sigma[n(2)]$	56.2242	34.4135	29.3493	34.3918
$E[c_1(2)]$	21.0623	46.2477	117.343	315.969
$\sigma[c_1(2)]$	5.13641	4.43681	5.61876	8.26569

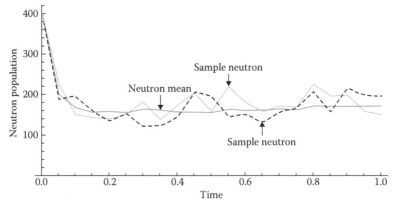

FIGURE 5.9
Neutron population density with $\alpha = 0.5$ and $\rho = -1/3$.

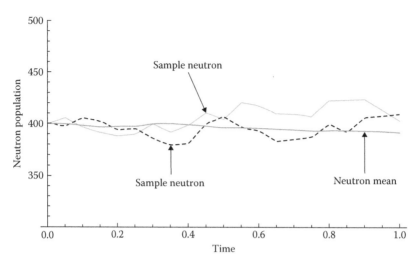

FIGURE 5.10
Neutron population density with $\alpha = 1$ and $\rho = -1/3$.

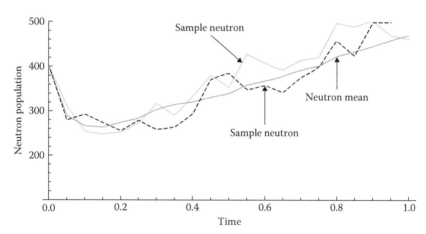

FIGURE 5.11
Neutron population density with $\alpha = 0.5$ and $\rho = 1/3$.

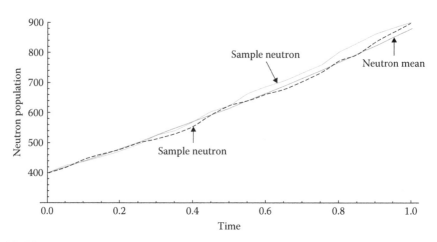

FIGURE 5.12
Neutron population density with $\alpha = 1$ and $\rho = 1/3$.

In the second example, we assume the initial condition as

$$\vec{x}(0) = 100 \begin{bmatrix} 1 \\ \dfrac{\beta_1}{\lambda_1 l} \\ \dfrac{\beta_2}{\lambda_2 l} \\ \vdots \\ \dfrac{\beta_m}{\lambda_m l} \end{bmatrix}$$

The parameters used in this example are [53,60] $\beta = 0.007$, $v = 2.5$, $l = 0.00002$ s, $q = 0$, $\rho = 0.003$, $\lambda_i = [0.0127, 0.0317, 0.115, 0.311, 1.4, 3.87]$, and $\beta_i = [0.000266, 0.001491, 0.001316, 0.002849, 0.000896, 0.000182]$ with $m = 6$ delayed groups. The computational results at $t = 0.1$ is given in Table 5.6, and the numerical results are presented graphically by Figures 5.13 through 5.20.

TABLE 5.6

Mean and Standard Deviation of Neutron and Precursor for Different Values of α with $\rho = 0.003$

	$\alpha = 0.25$	$\alpha = 0.5$	$\alpha = 0.75$	$\alpha = 1$ (Classical)
$E[n(0.1)]$	1.12011×10^{20}	6635.54	356.469	208.599
$\sigma[n(0.1)]$	4.45894×10^{20}	7755.62	664.685	255.954
$E[c_1(0.1)]$	5.50437×10^{22}	492804	164494	449815
$\sigma[c_1(0.1)]$	4.00827×10^{22}	112512	4794.21	1233.38

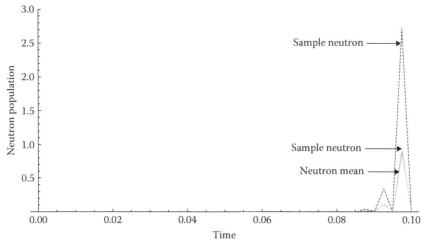

FIGURE 5.13
Neutron population density with $\alpha = 0.25$.

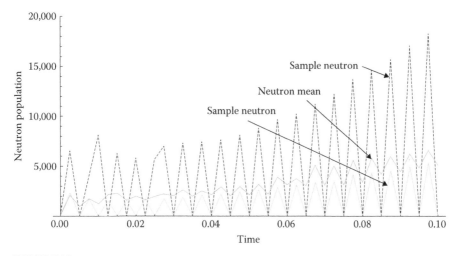

FIGURE 5.14
Neutron population density with $\alpha = 0.5$.

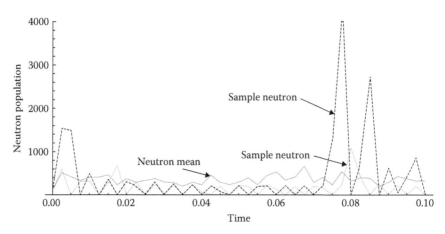

FIGURE 5.15
Neutron population density with $\alpha = 0.75$.

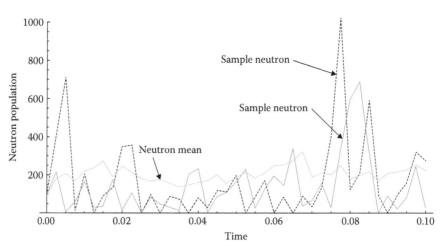

FIGURE 5.16
Neutron population density with $\alpha = 1$.

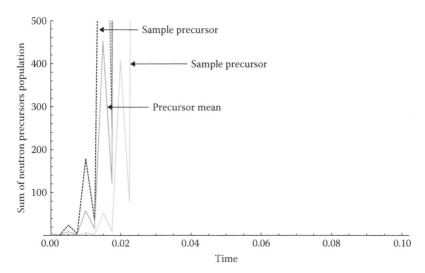

FIGURE 5.17

Sum of neutron precursors population with $\alpha = 0.25$.

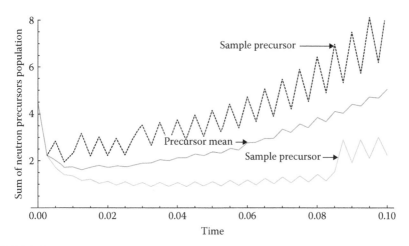

FIGURE 5.18

Sum of neutron precursors population with $\alpha = 0.5$.

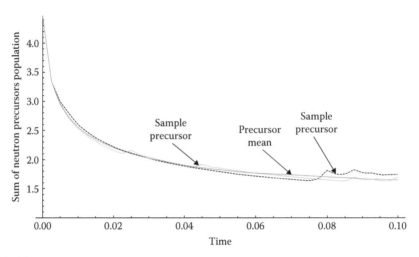

FIGURE 5.19
Sum of neutron precursors population with $\alpha = 0.75$.

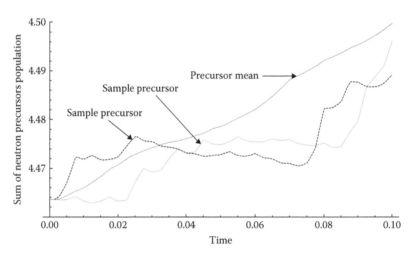

FIGURE 5.20
Sum of neutron precursors population with $\alpha = 1$.

5.8 Analysis for Stability of Numerical Computation for the FSNPK Equations

The stability of the numerical computation is calculated by taking the different time step size h with different values of the anomalous diffusion order α. In order to obtain a stable result, a set of time step sizes are considered by trial and error for different values of α.

A series of numerical experiments have been done for getting a better solution for the FSNPK equations. In the present numerical study, we consider $\alpha = 0.25$, 0.5, 0.75, and 1. The values of time step size h are taken between $0.005 \leq h \leq 0.025$. The simulation time was considered as 0.1 s. The stability criterion in this analysis is related with 0.01% of the relative error to $n_0 = 100$ at time 0.1 s for simulation.

$$0.01\% \geq \left| \frac{n_f - n_0}{n_f} \right| \times 100$$

where:
n_f is the neutron density that is calculated from fractional stochastic model

To exhibit the behavior of neutron density with $\alpha = 0.25$, we consider h in the interval [0.005 s, 0.025 s]. In order to check stability criteria, the relative error is 0.009403% at $h = 0.005$ s, 0.009928% at $h = 0.01$ s, and 0.0099998% at $h = 0.025$ s. To examine the behavior of neutron density with $\alpha = 0.5$, we consider h in the interval [0.005 s, 0.025 s] where the numerical solutions are similar with $\alpha = 0.25$. In this case, the relative error is 0.0088019% at $h = 0.005$ s, 0.008984% at $h = 0.01$ s, and 0.009141% at $h = 0.025$ s. The neutron density behavior can also be shown with $\alpha = 0.75$ by considering h in the interval [0.005 s, 0.025 s]. For this case, the numerical scheme is also stable for time step size $h = 0.005$ s whose corresponding relative error is 0.006554%, whereas the relative error is 0.006393% at $h = 0.01$ s and 0.006204% at $h = 0.025$ s. For the case when $\alpha = 1$ and h in the interval [0.005 s, 0.025 s], the relative error is 0.004932% at $h = 0.005$ s, 0.005230% at $h = 0.01$ s, and 0.002776% at $h = 0.025$ s. Therefore, the above numerical experiment confirms the stability of our numerical scheme for the solution of the FSNPK equations.

5.9 Conclusion

The numerical methods like the Euler–Maruyama method and strong order 1.5 Taylor method are clearly efficient and convenient for solving classical order stochastic neutron point kinetic equations [73,74]. The methods are easily applicable to obtain the solution of stochastic neutron point kinetic equations with sinusoidal and pulse reactivity functions. The obtained results exhibit its justification. This chapter shows the applicability of the two numerical stochastic methods like the Euler–Maruyama method and strong order 1.5 Taylor method for the numerical solution of stochastic neutron point kinetic equation with sinusoidal reactivity and pulse reactivity for one precursor. These methods are quite easy to apply for obtaining accurate numerical solutions for time-varying reactivities like sinusoidal and pulse reactivities. Moreover, the accuracy can be further improved when smaller the step size of each subinterval.

In this research work, the FSNPK equation also has been solved by using explicit finite difference method [75]. The method in this investigation clearly explores an effective numerical method for solving the FSNPK equation. The method is simple, efficient to calculate, and accurate with fewer round-off errors. This method can be used as powerful solvers for the FSNPK equation. The random behaviors of neutron density and neutron precursor concentrations have not been analyzed in fractional order prior to this research work. The results of the numerical approximations for the solution of neutron population density and sum of precursors population are also given graphically for different arbitrary values of α.

6

Solution for Nonlinear Classical and Fractional Order Neutron Point Kinetic Model with Newtonian Temperature Feedback Reactivity

6.1 Introduction

The multigroup delayed neutron point kinetic equations in the presence of temperature feedback reactivity are a system of stiff nonlinear ordinary differential equations. The neutron flux and the delayed neutron precursor concentration are important parameters for the study of safety and transient behavior of the reactor power. The point kinetic equations of multigroup delayed neutrons with temperature feedback reactivity describe the neutron density representing the reactor power level, time-dependent reactivity, precursor concentrations of multigroup delayed neutrons, and thermodynamic variables that enter into the reactivity equation. The solution of this system of equation is useful for providing an estimation for transient behavior of a reactor power and other systems of variables of the reactor cores that are fairly tightly coupled.

Fission neutrons are usually of different energies and move in different directions than the incident neutrons. Furthermore, there will generally be a change in the position, energy, and direction of motion of the neutron. The interactions of neutrons with nuclei in a medium thus result in transfer of the neutrons from one location to another, from one energy level to another, and from one direction to another. Thus, neutron population distribution in a nuclear reactor is described by transport equations. One of the simplest approximations to neutron transport that has been widely used in research and practice is the approximation given by the diffusion theory.

Reactivity is the most important parameter in nuclear reactor operation. When it is positive, the reactor is supercritical, zero at criticality; when it is negative, the reactor is subcritical. Reactivity can be controlled in various ways: by adding or removing fuel; by changing the fraction of neutrons that leaks from the system; or by changing the amount of an absorber that competes with the fuel for neutrons. The amount of reactivity in a reactor core determines the change in neutron population and reactor power. Hence, reactivity plays a vital role in reactor control.

In this research work, the explicit finite difference method (EFDM) [13,50,75] has been applied for solving the classical order and fractional order point kinetic equations of multigroup delayed neutrons with temperature feedback reactivity. The fractional neutron point kinetic equation is useful in the context of anomalous diffusion phenomena due to the highly heterogeneous configuration in nuclear reactors. Espinosa-Paredes et al. [49] proposed a fractional diffusion model as a constitutive equation of the neutron current density. This fractional diffusion model can be applied where large variations of neutron cross sections normally prevent the use of classical diffusion equation, specifically the presence of strong neutron absorbers in the fuel, control rods, and the coolant when injected boron forces the reactor to shutdown. Espinosa-Paredes et al. [49] proposed a solution procedure that has been inherited from Edwards et al. [54]. In contrast to their method, in the present numerical scheme fractional derivative has been discretized by Grünwald–Letnikov derivative, and the fractional point kinetic equation has been converted directly into finite difference equation. Then, it has been adjusted in the form of explicit finite difference scheme. The present numerical scheme is simple and efficient compared to the procedure proposed in ref. [49]. Although the Caputo derivative has been chosen by Espinosa-Paredes et al. [49], we have to apply the Grünwald–Letnikov fractional derivative instead of the Caputo derivative in the numerical computation for discretizing the fractional derivative. It is very much clear that the present numerical scheme requires less computational effort compared to generalization of the analytical exponential model (GAEM) [76] and Padé approximation method [77] for classical order $\alpha = 1$.

Due to the use of this finite difference numerical technique for solving coupled reactor kinetic equations, we apply the Grünwald–Letnikov fractional derivative, which is useful for discretizing the fractional derivative. The obtained results are presented graphically and also compared to the other methods that exist in open literature for classical order at $\alpha = 1$ [76,77].

6.2 Classical Order Nonlinear Neutron Point Kinetic Model

The multigroup delayed neutron point kinetic equations [22,23,32] and the Newtonian temperature feedback reactivity are the stiff nonlinear ordinary differential equations, which are presented as

$$\frac{dn(t)}{dt} = \left[\frac{\rho(t) - \beta}{l} \right] n(t) + \sum_{i=1}^{m} \lambda_i c_i(t) \tag{6.1}$$

$$\frac{dc_i(t)}{dt} = \left(\frac{\beta_i}{l}\right)n(t) - \lambda_i c_i(t), \quad i = 1, 2, 3, \ldots, m \tag{6.2}$$

and

$$\rho(t) = at - b\int_0^t n(t')dt' \tag{6.3}$$

where:
  $n(t)$ is the neutron density
  $\rho(t)$ is the reactivity as a function of time
  $\beta = \sum_{i=1}^{m}\beta_i$ is the total fraction of delayed neutrons
  β_i is the fraction of ith group of delayed neutrons
  λ_i is the decay constant of ith group of delayed neutrons
  l is the prompt neutron generation time
  $c_i(t)$ is the precursor concentrations of ith group of delayed neutron
  m is the total number of delayed neutrons group
  at is the impressed reactivity variation
  b is the shutdown coefficient of the reactor system

These equations have been the subject of extensive study using all kinds of approximations. We are interested here in the general problem, a time-dependent reactivity function $\rho(t)$. It is well known that these sets of ordinary differential equations are quite stiff.

One of the important factors in a nuclear reactor is reactivity, due to the fact that it is directly related to the control of the reactor. For safety analysis and analysis of the transient behavior of the reactor, the neutron population and the delayed neutron precursor concentration are important parameters to be studied. The start-up process of a nuclear reactor requires that reactivity is varied in the system by lifting the control rods discontinuously. In practice, the control rods are withdrawn at time intervals such that the reactivity is introduced in the reactor core linearly to allow criticality to be reached in a slow and safe manner.

6.3 Numerical Solution of Nonlinear Neutron Point Kinetic Equation in the Presence of Reactivity Function

Let us take the time step size h. Using the finite difference for time derivative, the numerical approximations of Equations 6.1 through 6.3, in view of the research work [13,50,75], have been obtained as

$$n_{k+1} = n_k + h\left(\frac{\rho_k - \beta}{l}\right)n_k + h\left(\sum_{i=1}^{m} \lambda_i c_{i,k}\right) \tag{6.4}$$

$$c_{i,k+1} = c_{i,k} + h\left[\left(\frac{\beta_i}{l}\right)n_k - \lambda_i c_{i,k}\right] \tag{6.5}$$

$$\rho_{k+1} = \rho_k + h(a - bn_k) \tag{6.6}$$

where:

$n_k = n(t_k)$ is neutron density at time t_k

$c_{i,k} = c_i(t_k)$ is the precursor concentrations of ith group of delayed neutron at time t_k

$\vec{x}_k = \vec{x}(t_k)$ is the kth approximation at time t_k, $t_k = kh$, $k = 0,1,2,....$ Here, $h = t_{k+1} - t_k$ with initial condition $\vec{x}_0 = \vec{x}(t_0)$

Equations 6.4 through 6.6 represent the explicit finite difference scheme that leads from the time layer t_k to t_{k+1}, where $k = 0,1,2,\cdots$. The main advantage of EFDM is that the method is relatively simple and easily computable.

6.4 Numerical Results and Discussions for the Classical Order Nonlinear Neutron Point Kinetic Equation

In the present analysis, we discuss three cases of reactivity function [76,78,79]: step, ramp (positive and negative), and feedback reactivities. All results started from equilibrium conditions with neutron density $n(0) = 1.0$ and kth of delayed neutron precursors density $c_k(0) = [n(0)\beta_k/\lambda_k l]$. In the following sections, each case will be discussed separately. The numerical results are presented in Tables 6.1 through 6.4 and are also illustrated graphically.

6.4.1 Step Reactivity Insertions

In this case, for checking the efficiency of the numerical scheme, it is applied to the thermal reactor with the following parameters [78,79]: $\lambda_1 = 0.0127 \text{ s}^{-1}$, $\lambda_2 = 0.0317 \text{ s}^{-1}$, $\lambda_3 = 0.115 \text{ s}^{-1}$, $\lambda_4 = 0.311 \text{ s}^{-1}$, $\lambda_5 = 1.40 \text{ s}^{-1}$, $\lambda_6 = 3.87 \text{ s}^{-1}$, $\beta_1 = 0.000285$, $\beta_2 = 0.0015975$, $\beta_3 = 0.00141$, $\beta_4 = 0.0030525$, $\beta_5 = 0.00096$, $\beta_6 = 0.000195$, $\beta = 0.0075$, and $l = 0.0005 \text{ s}$. The relative errors of neutron density with four cases of step reactivity $-1.0\$$, $-0.5\$$, $+0.5\$$, and $+1.0\$$ are presented in Table 6.1. The present scheme is an efficient numerical technique to obtain the solution for point kinetic equations with the step reactivity insertions.

TABLE 6.1

The Relative Errors and the Exact Neutron Density $n(t)$ (neutrons/cm³) of the Thermal Reactor with Step Reactivity for Step Size $h = 0.0001$ s

Reactivity ($)	Time (s)	Exact Solution [79]	Present Numerical Scheme Explicit Finite Difference Method	Relative Errors
−1.0	0.1	0.5205643	0.520454	2.11×10^{-4}
	1.0	0.4333335	0.433332	0
	10	0.2361107	0.23611	0
−0.5	0.1	0.6989252	0.698838	1.24×10^{-4}
	1.0	0.6070536	0.607053	0
	10	0.3960777	0.396077	0
+0.5	0.1	1.533113	1.53323	0
	1.0	2.511494	2.5115	0
	10	14.21503	14.2148	1.61×10^{-5}
+1.0	0.1	2.515766	2.51572	0
	0.5	10.36253	10.3614	0
	1.0	32.18354	32.176	2.34×10^{-4}

TABLE 6.2

The Neutron Density $n(t)$ (neutrons/cm³) of the Thermal Reactors with a Positive Ramp Reactivity (+0.1$/s) for Step Size $h = 0.0001$ s

Time (s)	TSM [79]	BBF [80]	SCM [81]	Explicit Finite Difference Method
2.0	1.3382	1.3382	1.3382	1.3382
4.0	2.2284	2.2284	2.2284	2.22842
6.0	5.5822	5.5820	5.5819	5.58192
8.0	42.789	42.786	42.788	42.7817
10.0	4.5143×10^{5}	4.5041×10^{5}	4.5391×10^{5}	4.49850×10^{5}

TABLE 6.3

The Neutron Density $n(t)$ (neutrons/cm³) of the Thermal Reactors with a Negative Ramp Reactivity (−0.1$/s) for Step Size $h = 0.0001$ s

Time (s)	TSM [79]	GAEM [76]	Padé [77]	Explicit Finite Difference Method
2.0	0.791955	0.792007	0.792007	0.792006
4.0	0.612976	0.613020	0.613018	0.613017
6.0	0.474027	0.474065	0.474058	0.474058
8.0	0.369145	0.369172	0.369169	0.369168
10.0	0.290636	0.290653	0.290654	0.290654

TABLE 6.4

The Neutron Density $n(t)$ (neutrons/cm³) at the First Peak of ²³⁵U reactor with Feedback Reactivity for Step Size $h = 0.0001$ s

a	b	Delayed Neutrons at the First Peak			Time of the First Peak		
		GAEM [76]	Padé [77]	Explicit Finite Difference Method	GAEM [76]	Padé [77]	Explicit Finite Difference Method
0.1	10^{-11}	2.4202×10^{11}	2.4197×10^{11}	2.42275×10^{11}	0.224	0.224	0.2245
	10^{-13}	2.9057×10^{13}	2.9055×10^{13}	2.9037×10^{13}	0.238	0.238	0.2385
0.01	10^{-11}	2.0103×10^{10}	2.0107×10^{10}	2.02264×10^{10}	1.100	1.100	1.1014
	10^{-13}	2.4882×10^{12}	2.4890×10^{12}	2.5043×10^{12}	1.149	1.149	1.150
0.003	10^{-11}	Not available	Not available	5.24943×10^{9}	Not available	Not available	2.897
	10^{-13}			6.73756×10^{11}			2.996

6.4.2 Ramp Reactivity Insertions

In this example, the numerical scheme is applied to the thermal reactor with the following parameters [78,79]: $\lambda_1 = 0.0127 \text{ s}^{-1}, \lambda_2 = 0.0317 \text{ s}^{-1}, \lambda_3 = 0.115 \text{ s}^{-1}$, $\lambda_4 = 0.311 \text{ s}^{-1}$, $\lambda_5 = 1.40 \text{ s}^{-1}$, $\lambda_6 = 3.87 \text{ s}^{-1}$, $\beta_1 = 0.000266$, $\beta_2 = 0.001491$, $\beta_3 = 0.001316$, $\beta_4 = 0.002849$, $\beta_5 = 0.000896$, $\beta_6 = 0.000182$, $\beta = 0.007$, and $l = 0.00002 \text{ s}$. The neutron density of the thermal reactor with positive ramp reactivity is $\rho(t) = +0.1t$ and the negative ramp reactivity is $\rho(t) = -0.1t$ whence the numerical results are provided in Tables 6.2 and 6.3, respectively.

6.4.3 Temperature Feedback Reactivity

In this case, the numerical scheme is applied to solve the point kinetic equations of delayed neutrons with the presence of Newtonian temperature feedback reactivity for ^{235}U–graphite reactor. The following parameters of ^{235}U–graphite reactor are used [76]: $\lambda_1 = 0.0127 \text{ sec}^{-1}, \lambda_2 = 0.0317 \text{ s}^{-1}$, $\lambda_3 = 0.115 \text{ s}^{-1}$, $\lambda_4 = 0.311 \text{ s}^{-1}$, $\lambda_5 = 1.40 \text{ s}^{-1}$, $\lambda_6 = 3.87 \text{ s}^{-1}$, $\beta_1 = 0.00246$, $\beta_2 = 0.001363$, $\beta_3 = 0.001203$, $\beta_4 = 0.002605$, $\beta_5 = 0.00819$, $\beta_6 = 0.00167$, and $\beta = 0.0064$. The value of generation time is 0.00005 s, a takes the values 0.1, 0.01, and 0.003, and b takes the values 10^{-11} and 10^{-13}. The numerical results obtained for feedback reactivity are presented in Table 6.4 and illustrated in Figures 6.1 through 6.6.

The objective of this study was to develop an accurate and computationally efficient method (explicit finite difference scheme) for solving time-dependent reactor dynamics equations with Newtonian temperature feedback. To test the developed solution and to prove the validity of the method for application purposes, a comparison with the other conventional methods indicates the superiority of the proposed EFDM.

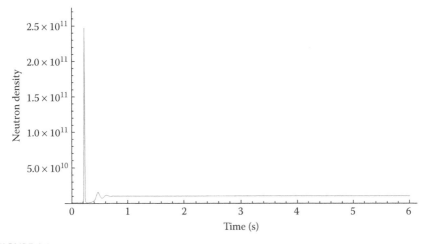

FIGURE 6.1
Neutron density $n(t)$ (neutrons/cm³) for $a = 0.1$ and $b = 10^{-11}$.

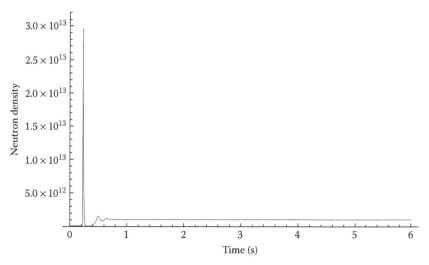

FIGURE 6.2
Neutron density $n(t)$ (neutrons/cm³) for $a = 0.1$ and $b = 10^{-13}$.

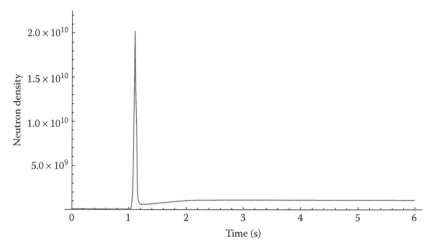

FIGURE 6.3
Neutron density $n(t)$ (neutrons/cm³) for $a = 0.01$ and $b = 10^{-11}$.

6.5 Mathematical Model for Nonlinear Fractional Neutron Point Kinetic Equation

The fractional calculus was first anticipated by Leibniz, one of the founders of standard calculus, in a letter written in 1695. Nowadays, real physical problems are best modeled by fractional calculus. This calculus involves different

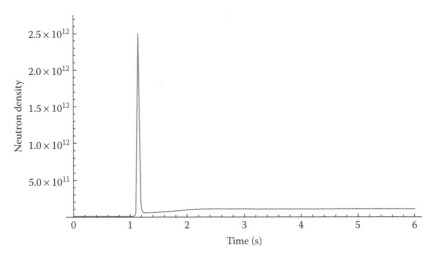

FIGURE 6.4
Neutron density $n(t)$ (neutrons/cm³) for $a = 0.01$ and $b = 10^{-13}$.

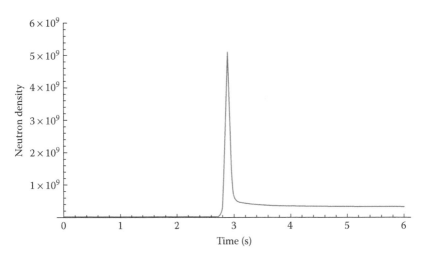

FIGURE 6.5
Neutron density $n(t)$ (neutrons/cm³) for $a = 0.003$ and $b = 10^{-11}$.

definitions of the fractional operators, namely, the Riemann–Liouville fractional derivative, Caputo derivative, Riesz derivative, and Grünwald–Letnikov fractional derivative [44,82]. The noninteger order of calculus has gained considerable importance during the past decades mainly due to its applications in diverse fields of science and engineering.

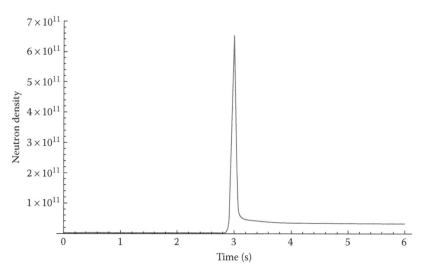

FIGURE 6.6
Neutron density $n(t)$ (neutrons/cm³) for $a = 0.003$ and $b = 10^{-13}$.

Here, we consider Grünwald–Letnikov definition of fractional derivative that is defined as

$$_aD_t^p f(t) = \lim_{\substack{h \to 0 \\ nh=t-a}} h^{-p} \sum_{r=0}^{n} \omega_r^p f(t - rh) \qquad (6.7)$$

where:

$$\omega_r^p = (-1)^r \binom{p}{r}$$

$$\omega_0^p = 1$$

$$\omega_r^p = \left(1 - \frac{p+1}{r}\right)\omega_{r-1}^p, \quad r = 1, 2, \cdots$$

Fractional neutron point kinetic equation is useful in the context of anomalous diffusion phenomena due to highly heterogeneous configuration in the nuclear reactors. The multigroup delayed fractional order ($\alpha > 0$) neutron point kinetic equations [83] with Newtonian temperature feedback reactivity are the stiff nonlinear ordinary differential equations [22–24, 32], which are presented as

$$\frac{d^\alpha n(t)}{dt^\alpha} = \left[\frac{\rho(t) - \beta}{l}\right]n(t) + \sum_{i=1}^{m} \lambda_i c_i(t) \qquad (6.8)$$

$$\frac{dc_i(t)}{dt} = \left(\frac{\beta_i}{l}\right) n(t) - \lambda_i c_i(t), \quad i = 1,2,3,\ldots,m \tag{6.9}$$

and

$$\rho(t) = at - b\int_0^t n(s)ds \tag{6.10}$$

where:
$n(t)$ is the neutron density
$\rho(t)$ is the reactivity as a function of time
$\beta = \sum_{i=1}^m \beta_i$ is the total fraction of delayed neutrons
β_i is the fraction of ith group of delayed neutrons
λ_i is the decay constant of ith group of delayed neutrons
l is the prompt neutron generation time
$c_i(t)$ is the precursor concentrations of ith group of delayed neutron
m is the total number of delayed neutrons group
at is the impressed reactivity variation
b is the shutdown coefficient of the reactor system

6.6 Application of EFDM for Solving the Fractional Order Nonlinear Neutron Point Kinetic Model

Let us take the time step size h. Using the finite difference for time derivative, the numerical approximations of the Equations 6.8 through 6.10, in view of the research work [13,50,75], have been obtained as

$$h^{-\alpha} \sum_{j=0}^r \omega_j^{(\alpha)} n_{(r-j)} = \left[\frac{\rho_{(r)} - \beta}{l}\right] n_{(r)} + \sum_{i=1}^m \lambda_i c_{i,r} \tag{6.11}$$

$$\left[\frac{c_{(r)} - c_{(r-1)}}{h}\right] = \left(\frac{\beta_i}{l}\right) n_{(r)} - \lambda_i c_{i,r}, \quad i = 1,2,3,\ldots,m \tag{6.12}$$

$$\left[\frac{\rho_{(r)} - \rho_{(r-1)}}{h}\right] = a - bn_{(r)} \tag{6.13}$$

Equations 6.11 through 6.13 lead to implicit numerical iteration scheme owing to the fact that EFDM is relatively simple and computationally fast. In this present work, we propose an explicit numerical scheme that leads from the time layer t_{r-1} to t_r as follows:

$$n_{(r)} = -\sum_{j=1}^{r} \omega_j^{(\alpha)} n_{(r-j)} + h^\alpha \left\{ \left[\frac{\rho_{(r-1)} - \beta}{l} \right] n_{r-1} + \sum_{i=1}^{m} \lambda_i c_{i,r-1} \right\} \tag{6.14}$$

where:

$n_{(r)} = n(t_r)$ is the rth approximation of neutron density at time t_r, with initial condition $n_0 = n(t_0)$

$c_{i,r} = c_i(t_r)$ is the precursor concentrations of ith group of delayed neutron at time t_r

$\rho_{(r)}$ is the rth approximation of feedback reactivity function

$t_r = rh, r = 0,1,2,...$

$\omega_j^\alpha = (-1)^j \binom{\alpha}{j}, j = 0,1,2,3,...$

$h = t_{r+1} - t_r$

The main advantage of EFDM is that the method is relatively simple and easily computable.

6.7 Numerical Results and Discussions for Fractional Nonlinear Neutron Point Kinetic Equation with Temperature Feedback Reactivity Function

In this case, the numerical scheme explicit finite difference is applied to solve the fractional order ($\alpha > 0$ and $\alpha \in R^+$) point kinetic equations with delayed neutrons in the presence of Newtonian temperature feedback reactivity for ^{235}U–graphite reactor. The Newtonian temperature feedback reactivity, which is dependent on time and neutron density, is given by $\rho(t) = at - b\int_0^t n(s)ds$, where the first term represents the impressed reactivity variation and b presents the shutdown coefficient of the reactor system. The point kinetic equation with temperature feedback corresponds to a stiff system of nonlinear differential equations for the neutron density and delayed precursor concentrations. The computed solutions of the present point kinetic equations provide information on the dynamics of a nuclear reactor operation in the presence of Newtonian temperature feedback reactivity. The following parameters of ^{235}U–graphite reactor are used [76,84]: $\lambda_1 = 0.0127\,s^{-1}$, $\lambda_2 = 0.0317\,s^{-1}$, $\lambda_3 = 0.115\,s^{-1}$, $\lambda_4 = 0.311\,s^{-1}$, $\lambda_5 = 1.40\,s^{-1}$, $\lambda_6 = 3.87\,s^{-1}$, $\beta_1 = 0.00246$, $\beta_2 = 0.001363$, $\beta_3 = 0.001203$, $\beta_4 = 0.002605$, $\beta_5 = 0.00819$, $\beta_6 = 0.00167$, and $\beta = 0.0064$. The value of generation time is $l = 5 \times 10^{-5}\,s$, a takes the values 0.1 and 0.01, and b takes the values 10^{-11} and 10^{-13}. We have considered the values for fractional order $\alpha = 0.5, 0.75, 1.25,$ and 1.5, respectively. If $\alpha = 1$, the process is normal diffusion, and when $0 < \alpha < 1$, then the process is anomalous diffusion. The fractional neutron point kinetic equation considering temperature feedback to reactivity with the Newtonian temperature approximation is analyzed in this present study. The numerical results obtained for neutron density of delayed neutrons in fractional order neutron point kinetic equation with feedback reactivity are introduced in Table 6.5 and with classical order $\alpha = 1$ in Table 6.6. The maximum peak for the neutron density can be observed from Figures 6.7 through 6.14 in the four cases of fractional order $\alpha = 0.5, 0.75, 1.25,$ and 1.75, using $a = 0.01$ in the presence of $b = 10^{-11}$ and $b = 10^{-13}$.

The behavior of the neutron density after the peak, illustrated in Figures 6.7 through 6.14, indicates that the system asymptotically leads to equilibrium state. In Figure 6.13, the second peak tries to get the equilibrium state and in Figure 6.12 after 1.6 s the system follows equilibrium state. From Figures 6.8 through 6.14, it can be observed that the neutron density of delayed neutron

TABLE 6.5

The Neutron Density $n(t)$ at the First Peak of ^{235}U reactor with Feedback Reactivity for Fractional Order Point Kinetic Equation When Step Size $h = 0.0001$ s

a	b	Delayed Neutron at the First Peak by Explicit Finite Difference Method $\alpha = 0.5, \alpha = 0.75, \alpha = 1.25, \alpha = 1.5$	Time of the First Peak by Explicit Finite Difference Method $\alpha = 0.5, \alpha = 0.75, \alpha = 1.25, \alpha = 1.5$
0.1	10^{-11}	$2.042 \times 10^{11}, 2.138 \times 10^{11}, 2.862 \times 10^{11}, 3.113 \times 10^{11}$	0.0919, 0.1391, 0.3549, 0.5144
	10^{-13}	$2.548 \times 10^{13}, 2.592 \times 10^{13}, 3.420 \times 10^{13}, 3.7511 \times 10^{13}$	0.0939, 0.1451, 0.3833, 0.5641
0.01	10^{-11}	$1.401 \times 10^{10}, 1.690 \times 10^{10}, 2.4605 \times 10^{10}, 2.8465 \times 10^{10}$	0.754, 0.8938, 1.377, 1.702
	10^{-13}	$1.7847 \times 10^{12}, 2.113 \times 10^{12}, 3.039 \times 10^{12}, 3.4975 \times 10^{12}$	0.763, 0.917, 1.4, 1.832

TABLE 6.6

The Neutron Density $n(t)$ (neutrons/cm^3) at the First Peak of ^{235}U reactor with Feedback Reactivity Taking Step Size $h = 0.0001$ s for Classical Order $\alpha = 1$

a	b	Delayed Neutrons at the First Peak			Time of the First Peak		
		GAEM [76]	Padé [77]	Explicit Finite Difference Method	GAEM [76]	Padé [77]	Explicit Finite Difference Method
0.1	10^{-11}	2.4202×10^{11}	2.4197×10^{11}	2.42275×10^{11}	0.224	0.224	0.2245
	10^{-13}	2.9057×10^{13}	2.9055×10^{13}	2.9037×10^{13}	0.238	0.238	0.2385
0.01	10^{-11}	2.0103×10^{10}	2.0107×10^{10}	2.02264×10^{10}	1.100	1.100	1.1014
	10^{-13}	2.4882×10^{12}	2.4890×10^{12}	2.5043×10^{12}	1.149	1.149	1.150

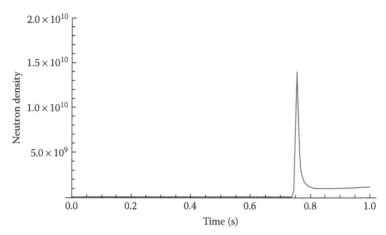

FIGURE 6.7
Neutron density of delayed neutrons for $\alpha = 0.5$ with $a = 0.01$ and $b = 10^{-11}$.

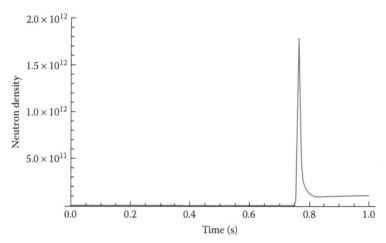

FIGURE 6.8
Neutron density of delayed neutrons for $\alpha = 0.5$ with $a = 0.01$ and $b = 10^{-13}$.

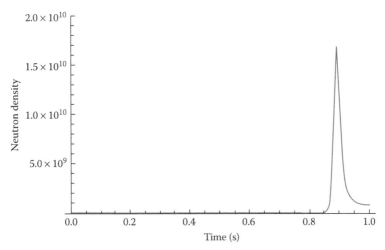

FIGURE 6.9
Neutron density of delayed neutrons for $\alpha = 0.75$ with $a = 0.01$ and $b = 10^{-11}$.

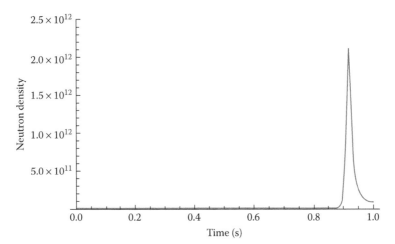

FIGURE 6.10
Neutron density of delayed neutrons for $\alpha = 0.75$ with $a = 0.01$ and $b = 10^{-13}$.

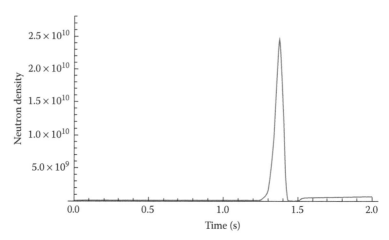

FIGURE 6.11
Neutron density of delayed neutrons for $\alpha = 1.25$ with $a = 0.01$ and $b = 10^{-11}$.

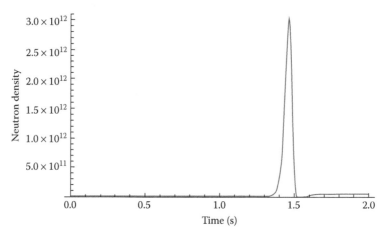

FIGURE 6.12
Neutron density of delayed neutrons for $\alpha = 1.25$ with $a = 0.01$ and $b = 10^{-13}$.

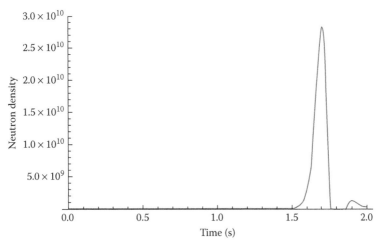

FIGURE 6.13

Neutron density of delayed neutrons for $\alpha = 1.5$ with $a = 0.01$ and $b = 10^{-11}$.

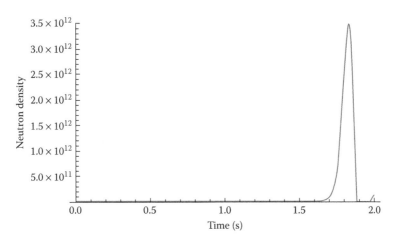

FIGURE 6.14

Neutron density of delayed neutrons for $\alpha = 1.5$ with $a = 0.01$ and $b = 10^{-13}$.

142   *Fractional Calculus with Applications for Nuclear Reactor Dynamics*

TABLE 6.7

Absolute Errors for Neutron Density $n(t)$ (neutrons/cm³) at the First Peak of ^{235}U reactor in the Presence of Feedback Reactivity Taking Step Size $h = 0.0001$ s for Classical Order $\alpha = 1$

		Delayed Neutrons at the First Peak		Time of the First Peak	
a	b	Absolute Error of Present Method (EFDM) with Regard to GAEM [76]	Absolute Error of Present Method (EFDM) with Regard to Padé [77]	Absolute Error of Present Method (EFDM) with Regard to GAEM [76]	Absolute Error of Present Method (EFDM) with Regard to Padé [77]
0.1	10^{-11}	0.00255×10^{11}	0.00305×10^{11}	0.0005	0.0005
	10^{-13}	0.002×10^{13}	0.0018×10^{13}	0.0005	0.0005
0.01	10^{-11}	0.01234×10^{10}	0.0119×10^{10}	0.0014	0.0014
	10^{-13}	0.0161×10^{12}	0.0153×10^{12}	0.001	0.001

tries to follow equilibrium behavior after a high density in the first peak. The obtained results from explicit finite difference scheme for neutron density using temperature feedback reactivity have been compared with the GAEM [76] and Padé approximation method [77] for classical order $\alpha = 1$. Moreover, the absolute errors of the proposed scheme with respect to the GAEM [76] and Padé approximation method [77] have been presented in Table 6.7. In this present research work, the behavior of the first peak of the neutron density in fractional order using temperature feedback reactivity has been analyzed. This is the first time this analysis has been accomplished.

6.8 Computational Error Analysis for the Fractional Order Nonlinear Neutron Point Kinetic Equation

Grünwald–Letnikov fractional derivative presented in Equation 6.7 is important for discretizing fractional derivative $d^\alpha n(t)/dt^\alpha$ numerically in a simple and efficient way:

$$\frac{d^\alpha n(t)}{dt^\alpha} = \frac{1}{h^\alpha} \sum_{j=0}^{r} \omega_j^{(\alpha)} n_{(r-j)} + O(h^p)$$

where:
p is the order of approximation

The truncation error in Equation 6.11 is

$$\frac{1}{h^\alpha} \sum_{j=0}^{r} \omega_j^{(\alpha)} n_{(r-j)} - \left[\frac{\rho_{(r)} - \beta}{l}\right] n_{(r)} - \sum_{i=1}^{m} \lambda_i c_{i,r} = O(h^p)$$

Again, the truncating error in Equation 6.12 is

$$\left[\frac{c_{(r)} - c_{(r-1)}}{h}\right] - \left(\frac{\beta_i}{l}\right) n_{(r)} + \lambda_i c_{i,r} = O(h)$$

And finally, the truncating error in Equation 6.13 is

$$\left[\frac{\rho_{(r)} - \rho_{(r-1)}}{h}\right] - a + bn_{(r)} = O(h)$$

Therefore, we finally get the total truncation error for Equations. 6.11 through 6.13 as follows:

$$= O(h^p) + O(h) + O(h) \cong O(h^p)$$

Hence, we obtain the computational error as $O(h^p)$ and eventually it tends to zero as $h \to 0$. Moreover, it shows that the computational overhead is less for the proposed explicit finite difference scheme.

6.9 Conclusion

The objective of this study is to develop an accurate and computationally efficient method for solving time-dependent reactor dynamics equations. The numerical solutions of integer and fractional order nonlinear point kinetic equations with multigroup of delayed neutrons are presented. The EFDM constitutes an easy algorithm that provides the results with sufficient accuracy for most applications and is being both conceptually and structurally simple. The method is simple, efficient to calculate, and accurate with fewer computational errors. Results of this method are compared with other available methods from which it can be concluded that the method is simple and computationally fast. To assess utility of the developed technique, different cases of reactivity were studied.

The numerical solutions for both classical and fractional order point kinetic equations [83] with multigroup of delayed neutrons are presented in the presence of Newtonian temperature feedback reactivity. To predict the dynamical behavior for ^{235}U reactors with time-dependent reactivity function and to obtain the solution of multigroup delayed neutron point kinetic equation [67,71], this numerical scheme (EFDM) is simple and efficient method. This numerical technique gives a good result for the nonlinear neutron point kinetic equations with time-dependent reactivity function. The method is simple, efficient to calculate, and accurate with fewer computational errors. The obtained results exhibit its justification. In Table 6.6, the solutions of classical order ($\alpha = 1$) point kinetic equations using Newtonian temperature feedback reactivity have been compared with the results available in the works of Nahla and Aboander and Nahla [76,77]. From Table 6.6, it can be observed that there is a very good agreement of results between the present method and the methods, namely, the GAEM and Padé approximation method, applied in the works of Nahla and Aboander and Nahla [76,77]. The authors have provided the computed absolute errors for their proposed numerical scheme with regard to the other reference results that exist in open literature in Table 6.7 for classical integer order (i.e., $\alpha = 1$). In this aspect, it can be concluded that the present method is a very simple and efficient technique. The present analysis exhibits that the featured method is applicable equally well to nonlinear problem in which the reactivity depends on the neutron density through temperature feedback. This research work shows the applicability of the EFDM for the numerical solution of nonlinear classical and fractional order neutron point kinetic equation. The results representing the numerical approximate solutions of neutron density involving feedback reactivity have been illustrated graphically.

7

Numerical Simulation Using Haar Wavelet Operational Method for Neutron Point Kinetic Equation Involving Imposed Reactivity Function

7.1 Introduction

The numerical solution of point kinetic equation with a group of delayed neutrons is useful in predicting neutron density variation during the operation of a nuclear reactor. The continuous indication of the neutron density and its rate of change is important for the safe start-up and operation of reactors. The Haar wavelet operational method (HWOM) has been proposed to obtain the numerical approximate solution of the neutron point kinetic equation that appears in a nuclear reactor with time-dependent and -independent reactivity functions. The present method has been applied to solve stiff point kinetic equations elegantly with step, ramp, zig-zag, sinusoidal, and pulse reactivity insertions. This numerical method has turned out as an accurate computational technique for many applications. In the dynamical system of a nuclear reactor, the point kinetic equations are the coupled linear differential equations for neutron density and delayed neutron precursor concentrations. These equations, which express the time dependence of the neutron population and the decay of the delayed neutron precursors within a reactor, are first order and linear, and essentially describe the change in neutron population within the reactor due to a change in reactivity. As reactivity is directly related to the control of the reactor, it is the important property in a nuclear reactor. For the purpose of safety analysis and transient behavior of the reactor, the neutron population and the delayed neutron precursor concentration are important parameters to be studied. An important property of the kinetic equations is the stiffness of the system. The stiffness is a severe problem in numerical solutions of the point kinetic equations, and it necessarily requires the need for small time steps in a computational scheme.

Aboanber and Nahla [77] presented an analytical Padé approximate solution for a group of six delayed neutrons with constant, ramp, and temperature feedback reactivity insertions. Again Aboanber and Nahla [85,86] presented the solution of a point kinetic equation with exponential mode analysis and generalization of analytical inversion method. Kinard and Allen [53] described the numerical solution based on piecewise constant approximation for the point kinetic equations in nuclear reactor dynamics. Nahla [76–79] presented the analytical methods to solve nonlinear point kinetic equations, and generalized power series solution for neutron point kinetic equation proposed by Hamada [87]. A numerical integral method that efficiently provides the solution of the point kinetic equations by using the better basis function (BBF) for the approximation of the neutron density in one-time step integrations has been described and investigated by Li et al. [80]. Chao and Attard [81] proposed the stiffness confinement method (SCM) for solving the kinetic equations to overcome the stiffness problem in reactor kinetics.

Quintero-Leyva [63] solved the neutron point kinetic equation by a numerical algorithm CORE for a lumped and temperature feedback. Using very simple technique like the backward Euler finite difference (BEFD) method, Ganapol [88] solved neutron point kinetic equation. McMohan and Pierson [64] solved the neutron point kinetic equation using the Taylor series method (TSM) involving reactivity functions. Again Picca et al. [89] solved the neutron point kinetic equation by applying enhanced piecewise constant approximation (EPCA) involving both linear and nonlinear reactivity insertion.

Wavelet analysis is a newly developed mathematical tool for applied analysis, image manipulation, and numerical analysis. Wavelets have been applied in numerous disciplines such as image compression, data compression, and many more [90,91]. Among the different wavelet families, mathematically most simple are the Haar wavelets [92].

Haar functions have been used from 1910 when they were introduced by the Hungarian mathematician Alfred Haar [93]. The Haar transform is one of the earliest examples of what is known now as a compact, dyadic, orthonormal wavelet transform. The Haar function, being an odd rectangular pulse pair, is the simplest and oldest orthonormal wavelet with compact support. In the meantime, several definitions of the Haar functions and various generalizations have been published and used. They were intended to adapt this concept to some practical applications, as well as to extend its application to different classes of signals. Thanks to their useful features and possibility to provide a local analysis of signals, the Haar functions appear very attractive in many applications, for example, image coding, edge extraction, and binary logic design.

Haar wavelets are made up of pairs of piecewise constant functions and mathematically the simplest orthonormal wavelets with a compact support. Due to the mathematical simplicity, the Haar wavelets method has turned out to be an effective tool for solving differential and integral equations. The Haar wavelets have the following features: (1) orthogonal and normalization, (2) having compact support, and (3) the simple expression [94,95]. Due to its simplicity, the Haar wavelets are very effective for solving differential and integral equations. It is a worthwhile attempt to develop the numerical scheme for the solution of the point reactor kinetic equations. The importance of this scheme is that it can usually be applied to more realistic mathematical or physical models. Therefore, the main focus of the present analysis is the application of Haar wavelet technique for solving the problem of coupled point kinetic equations with reactivity function in nuclear reactor dynamics.

In this present work, the HWOM for solving the point kinetic equation with a group of six delayed neutrons has been applied. The obtained numerical approximation results of this method are then compared with the referenced methods for different reactivities. In the present investigation, the main attractive advantage of this computational numerical method is its simplicity, efficiency, and applicability.

7.2 Haar Wavelets

Haar wavelets are the simplest wavelets among various types of wavelets. They are step functions on the real line that can take only three values −1, 0, and 1. The method has been used for its simplicity and computationally fast attractive features. The Haar functions are the family of switched rectangular waveforms where amplitudes can differ from one function to another function. Usually, the Haar wavelets are defined for the interval $t \in [0, 1]$ but in general case $t \in [A, B]$, we divide the interval $[A, B]$ into m equal subintervals, each of width $\Delta t = (B - A)/m$. In this case, the orthogonal set of Haar functions is defined in the interval $[A, B]$ by [95]

$$h_0(t) = \begin{cases} 1 & t \in [A, B], \\ 0 & \text{elsewhere,} \end{cases} \quad \text{and } h_i(t) = \begin{cases} 1, & \zeta_1(i) \le t < \zeta_2(i) \\ -1, & \zeta_2(i) \le t < \zeta_3(i) \\ 0, & \text{otherwise} \end{cases} \quad (7.1)$$

where:

$$\zeta_1(i) = A + \left(\frac{k-1}{2^j}\right)(B - A) = A + \left(\frac{k-1}{2^j}\right)m\Delta t$$

$$\zeta_2(i) = A + \left[\frac{k-(1/2)}{2^j}\right](B - A) = A + \left[\frac{k-(1/2)}{2^j}\right]m\Delta t$$

$$\zeta_3(i) = A + \left(\frac{k}{2^j}\right)(B - A) = A + \left(\frac{k}{2^j}\right)m\Delta t$$

$i = 1, 2, \cdots, m$, $m = 2^J$, and J is a positive integer, called the maximum level of resolution. Here, j and k represent the integer decomposition of the index i, that is, $i = k + 2^j - 1$, $0 \le j < i$, and $1 \le k < 2^j + 1$.

7.3 Function Approximation and Operational Matrix of the General Order Integration

Any function $y(t) \in L^2([0, 1])$ can be expanded in Haar series as

$$y(t) = c_0 h_0(t) + c_1 h_1(t) + c_2 h_2(t) + \cdots, \quad \text{where } c_j = \int_0^1 y(t) h_j(t) dt \qquad (7.2)$$

If $y(t)$ is approximated as piecewise constant in each subinterval, the sum in Equation 7.2 may be terminated after m terms and consequently we can write discrete version in the matrix form as

$$Y \approx \sum_{i=0}^{m-1} c_i h_i(t_l) = C_m^T H_m \qquad (7.3)$$

for collocation points

$$t_l = A + (l - 0.5)\Delta t, \quad l = 1, 2, \cdots, m \qquad (7.4)$$

where:
   Y and C_m^T are the m-dimensional row vectors

Here H is the Haar wavelet matrix of order m defined by $H = [h_0, h_1, \cdots, h_{m-1}]^T$, that is,

$$H = \begin{bmatrix} h_0 \\ h_1 \\ \vdots \\ h_{m-1} \end{bmatrix} = \begin{bmatrix} h_{0,0} & h_{0,1} & \cdots & h_{0,m-1} \\ h_{1,0} & h_{1,1} & \cdots & h_{1,m-1} \\ \vdots & & & \\ h_{m-1,0} & h_{m-1,1} & \cdots & h_{m-1,m-1} \end{bmatrix} \qquad (7.5)$$

where:
   $h_0, h_1, \cdots, h_{m-1}$ are the discrete form of the Haar wavelet bases

7.3.1 Operational Matrix of the General Order Integration

The integration of the $H_m(t) = [h_0(t), h_1(t), \ldots, h_{m-1}(t)]^T$ can be approximated by Chen and Hsiao [96]

$$\int_0^t H_m(\tau) d\tau \cong Q H_m(t) \qquad (7.6)$$

where Q is called as Haar wavelet operational matrix of integration, which is a square matrix of m dimension. To derive the Haar wavelet operational matrix of the general order of integration, we recall the fractional integral of order $\alpha(> 0)$ that is defined by Podlubny [44]

$$J^\alpha f(t) = \frac{1}{\Gamma(\alpha)} \int_0^t (t-\tau)^{\alpha-1} f(\tau) d\tau, \, t > 0, \, \alpha \in R^+ \tag{7.7}$$

where:
 R^+ is the set of positive real numbers

The Haar wavelet operational matrix Q^α for integration of the general order α is given by

$$Q^\alpha H_m(t) = J^\alpha H_m(t) = \left[J^\alpha h_0(t), J^\alpha h_1(t), \cdots, J^\alpha h_{m-1}(t) \right]^T$$
$$= \left[Qh_0(t), Qh_1(t), \cdots, Qh_{m-1}(t) \right]^T \tag{7.8}$$

where:

$$Qh_0(t) = \begin{cases} \dfrac{t^\alpha}{\Gamma(1+\alpha)}, & t \in [A, B] \\ 0, & elsewhere \end{cases} \tag{7.9}$$

and

$$Qh_i(t) = \begin{cases} 0, & A \le t < \zeta_1(i) \\ \phi_1, & \zeta_1(i) \le t < \zeta_2(i) \\ \phi_2, & \zeta_2(i) \le t < \zeta_3(i) \\ \phi_3, & \zeta_3(i) \le t < B \end{cases} \tag{7.10}$$

where:

$$\phi_1 = \frac{[t - \zeta_1(i)]^\alpha}{\Gamma(\alpha+1)}$$

$$\phi_2 = \frac{[t - \zeta_1(i)]^\alpha}{\Gamma(\alpha+1)} - 2 \frac{[t - \zeta_2(i)]^\alpha}{\Gamma(\alpha+1)}$$

$$\phi_3 = \frac{[t - \zeta_1(i)]^\alpha}{\Gamma(\alpha+1)} - 2 \frac{[t - \zeta_2(i)]^\alpha}{\Gamma(\alpha+1)} + \frac{[t - \zeta_3(i)]^\alpha}{\Gamma(\alpha+1)}$$

for $i = 1, 2, \cdots, m$, $m = 2^J$, and J, a positive integer, is called the maximum level of resolution. Here, j and k represent the integer decomposition of the index i, that is, $i = k + 2^j - 1$, $0 \le j < i$, and $1 \le k < 2^j + 1$.

7.4 Application of the HWOM for Solving Neutron Point Kinetic Equation

The point reactor kinetic equations without effect of the extraneous neutron source are given by [24,32]

$$\frac{dn(t)}{dt} = \left[\frac{\rho(t) - \beta}{l}\right] n(t) + \sum_{i=1}^{M} \lambda_i c_i \tag{7.11}$$

$$\frac{dc_i(t)}{dt} = \frac{\beta_i}{l} n(t) - \lambda_i c_i(t), \quad i = 1, 2, \ldots, M \tag{7.12}$$

where:
$n(t)$ is the time-dependent neutron density
$c_i(t)$ is the ith precursor density
$\rho(t)$ is the time-dependent reactivity function
β_i is the ith delayed fraction
$\beta = \sum_{i=1}^{M} \beta_i$ is the total delayed fraction
l is the neutron generation time
λ_i is the ith group decay constant

with the initial condition $n(0) = n_0$ and $c_k(0) = \left[n(0)\beta_k / \lambda_k l\right]$, for $k = 1, 2, \ldots, M$.

Let us divide the interval $[A, B]$ into m equal subintervals, each of width $\Delta t = (B - A)/m$. We assume

$$\frac{dn(t)}{dt} = Dn(t) = \sum_{i=1}^{m} c_i h_i(t) \tag{7.13}$$

Integrating Equation 7.13 from 0 to t, we can derive that

$$n(t) = n(0) + \sum_{i=1}^{m} c_i Qh_i(t) \tag{7.14}$$

where $Qh_i(t)$ is given by Equation 7.10 for $\alpha = 1$.

Next, we consider

$$\frac{dc_k(t)}{dt} = Dc_k(t) = \sum_{i=1}^{m} d_{k,i} h_i(t) \tag{7.15}$$

From Equation 7.15, we can again derive that

$$c_k(t) = c_k(0) + \sum_{i=1}^{m} d_{k,i} Qh_i(t) \tag{7.16}$$

Here also $Qh_i(t)$ is given by Equation 7.10 for $\alpha = 1$.

Now, we substitute all these Equations 7.13 through 7.16 together in Equations 7.11 and 7.12 to obtain the following numerical scheme as [97]

$$\sum_{i=1}^{m} c_i h_i(t) = \frac{(\rho - \beta)}{l} \left[n(0) + \sum_{i=1}^{m} c_i Q h_i(t) \right] + \sum_{r=1}^{M} \lambda_r \left[c_r(0) + \sum_{i=1}^{m} d_{r,i} Q h_i(t) \right] \quad (7.17)$$

and

$$\sum_{i=1}^{m} d_{k,i} h_i(t) = \left(\frac{\beta_k}{l} \right) \left[n(0) + \sum_{i=1}^{m} c_i Q h_i(t) \right] - \lambda_k \left[c_k(0) + \sum_{i=1}^{m} d_{k,i} Q h_i(t) \right] \quad (7.18)$$

By considering all the wavelet collocation points $t_l = A + (l - 0.5)\Delta t, l = 1, 2, \cdots, m$, we may obtain the system of m number of algebraic equations. By solving these systems of equations using any mathematical software, we can obtain the Haar coefficients c_i and $d_{k,i}$. Then, we can find the approximate solution for neutron density $n(t)$. The main advantage of the HWOM is that it always converts the system of equations into a set of algebraic equations that can be easily solvable by any mathematical software. The limitation of this wavelet method is only that it requires a high level of resolution for large time duration.

7.5 Numerical Results and Discussions

In this chapter, the HWOM has been used for numerical solution of the point reactor kinetic equations (Equations 7.11 and 7.12). In the present analysis, the above numerical method has been applied to solve the point kinetic equations with a group of six delayed neutrons. The four types of cases involving step, ramp, zig-zag, and sinusoidal reactivates have been presented. The values for λ_k and β_k and neutron generation time l for the reactor have been taken from Ganapol [88]. All results started from equilibrium conditions with neutron density $n(0) = 1.0$ and kth of delayed neutron precursors density $c_k(0) = [n(0)\beta_k / \lambda_k l]$. In the subsequent sections, each case will be discussed separately.

7.5.1 Step Reactivity Insertions

To check the accuracy of the Haar wavelet method, it has been applied to the thermal reactor with the following parameters obtained from Ganapol [88]: $\lambda_1 = 0.0127 \text{ s}^{-1}$, $\lambda_2 = 0.0317 \text{ s}^{-1}$, $\lambda_3 = 0.115 \text{ s}^{-1}$, $\lambda_4 = 0.311 \text{ s}^{-1}$, $\lambda_5 = 1.40 \text{ s}^{-1}$, $\lambda_6 = 3.87 \text{ s}^{-1}$, $\beta = 0.0075$, $\beta_1 = 0.000285$, $\beta_2 = 0.0015975$, $\beta_3 = 0.00141$, $\beta_4 = 0.0030525$, $\beta_5 = 0.00096$, $\beta_6 = 0.000195$, and $l = 0.0005 \text{ s}$. The obtained results justify that the HWOM is an accurate and efficient computational technique for solving point kinetic equation using step reactivity insertions.

TABLE 7.1

Neutron Density $n(t)$ of the Thermal Reactor with Step
Reactivity Insertions

$\rho(\$)$	$t(s)$	HWOM with $m = 2^8$	BEFD [88]
−1.0	0.1	5.20563444767E−01	5.205642866E−01
	1.0	4.33333406022E−01	4.333334453E−01
	10	2.36110396447E−01	2.361106508E−01
	100	2.86673059802E−02	2.866764245E−02
−0.5	0.1	6.989247254605E−01	6.989252256E−01
	1.0	6.070535313626E−01	6.070535656E−01
	10	3.960774713631E−01	3.960776907E−01
	100	7.158236238460E−02	7.158285444E−02
+0.5	0.1	1.53311290367E+00	1.53311264E+00
	1.0	2.51149468921E+00	2.511494291E+00
	10	1.42151199462E+01	1.421502524E+01
	100	8.05861718287E+07	8.006143562E+07
+1.0	0.1	2.51576625250E+00	2.515766141E+00
	0.5	1.036255172721E+01	1.036253381E+01
	1.0	3.218390468938E+01	3.218354095E+01
	10	3. 282476942940E+09	3.246978898E+09

Table 7.1 represents the solution for neutron density involving step reactivity −1.0$, −0.5$, +0.5$, and +1.0$ using HWOM, and consequently, the obtained results have been compared with the BEFD method [88] taking level of resolution $J = 8$.

7.5.2 Ramp Reactivity Insertions

In this case, the Haar wavelet method has been applied to the thermal reactor with the following parameters obtained from Nahla [78,79] and Ganapol [88]: $\lambda_1 = 0.0127 \text{ s}^{-1}$, $\lambda_2 = 0.0317 \text{ s}^{-1}$, $\lambda_3 = 0.115 \text{ s}^{-1}$, $\lambda_4 = 0.311 \text{ s}^{-1}$, $\lambda_5 = 1.40 \text{ s}^{-1}$, $\lambda_6 = 3.87 \text{ s}^{-1}$, $\beta = 0.007$, $l = 0.00002 \text{ s}$, $\beta_1 = 0.000266$, $\beta_2 = 0.001491$, $\beta_3 = 0.001316$, $\beta_4 = 0.002849$, $\beta_5 = 0.000896$, and $\beta_6 = 0.000182$. Two cases of ramp reactivity, one is positive and another is negative ramp, are introduced.

7.5.2.1 Positive Ramp Reactivity

Here, the neutron density of the thermal reactor with positive ramp reactivity $\rho(t) = 0.1t$ is introduced. The numerical solution of neutron density with positive ramp reactivity obtained from HWOM with $m = 1024$ number of collocation points is given in Table 7.2. Also Table 7.2 represents the comparison results with the TSM [64], BBF [80], SCM [81], and with the BEFD [88]. It can be observed that there is a good agreement between the obtained results and the other available results. More accurate results for large time can be obtained with increase in collocation points.

TABLE 7.2

The Neutron Density $n(t)$ of the Thermal Reactor with Positive Ramp Reactivity (+0.1$/s)

Time (s)	TSM [64]	BBF [80]	SCM [81]	HWOM with $m = 1024$	BEFD [88]
2.0	1.3382E+00	1.3382E+00	1.3382E+00	1.338200011E+00	1.338200050E+00
4.0	2.2284E+00	2.2284E+00	2.2284E+00	2.228441300E+00	2.228441897E+00
6.0	5.5822E+00	5.5820E+00	5.5819E+00	5.582043255E+00	5.582052449E+00
8.0	4.2789E+01	4.2786E+01	4.2788E+01	4.278584501E+01	4.278629573E+01
10.0	4.5143E+05	4.5041E+05	4.5391E+05	4.522953180E+05	4.511636239E+05
11.0	NA	NA	NA	1.828233653E+16	1.792213607E+16

7.5.2.2 Negative Ramp Reactivity

The neutron density of the thermal reactor with a negative ramp reactivity, $\rho(t) = -0.1t$, has been exhibited in Table 7.3. Table 7.3 represents the solution for neutron density with negative ramp reactivity obtained from the HWOM and after comparison with the TSM [64], generalized analytical exponential method [76], and Padé approximation method [77] taking $m = 1024$ number of collocation points. Thus, it can be also observed that there is a good agreement between the obtained results and the other available results.

7.5.3 Zig-Zag Reactivity

In this case, a zig-zag reactivity function for the thermal reactor has been considered as follows:

$$\rho(t) = \begin{cases} 0.0075t, & 0 \leq t \leq 0.5 \\ -0.0075(t-0.5)+0.00375, & 0.5 \leq t \leq 1 \\ 0.0075(t-1), & 1 \leq t \leq 1.5 \\ 0.00375, & 1.5 \leq t \end{cases}$$

TABLE 7.3

The Neutron Density $n(t)$ of the Thermal Reactor with Negative Ramp Reactivity (−0.1$/s)

Time (s)	TSM [64]	GAEM [76]	Padé [77]	HWOM with $m = 1024$
2.0	7.91955E–01	7.92007E–01	7.92007E–01	7.920047444289586E–01
4.0	6.12976E–01	6.13020E–01	6.13018E–01	6.13015745981558E–01
6.0	4.74027E–01	4.74065E–01	4.74058E–01	4.74056567958244E–01
8.0	3.69145E–01	3.69172E–01	3.69169E–01	3.69167718232784E–01
10.0	2.90636E–01	2.90653E–01	2.90654E–01	2.90653117814878E–01
11.0	NA	NA	NA	2.59129648117922E–01

where the following kinetic parameters are taken from Picca et al. [89]: $\lambda_1 = 0.0127\,s^{-1}$, $\lambda_2 = 0.0317\,s^{-1}$, $\lambda_3 = 0.115\,s^{-1}$, $\lambda_4 = 0.311\,s^{-1}$, $\lambda_5 = 1.40\,s^{-1}$, $\lambda_6 = 3.87\,s^{-1}$, $l = 0.0005\,s$, $\beta_1 = 0.000285$, $\beta_2 = 0.0015975$, $\beta_3 = 0.001410$, $\beta_4 = 0.0030525$, $\beta_5 = 0.00096$, $\beta_6 = 0.000195$, and $\beta = 0.0075$.

Table 7.4 represents the numerical solution for neutron density involving zig-zag reactivity obtained from the HWOM by considering the level of resolution $J = 8$. Simultaneously, the obtained results have also been compared with EPCA [89].

7.5.4 Sinusoidal Reactivity Insertion

7.5.4.1 For Group of One Delayed Neutron

To check the accuracy of the Haar wavelet method with nonlinear reactivity insertion, it is applied to the fast reactor with the following parameters obtained from Ganapol [88]: $\lambda_1 = 0.077\,s^{-1}$, $\beta = \beta_1 = 0.0079$, $l = 10^{-7}\,s$, and $\rho_0 = 0.0053333$. The reactivity is a time-dependent function of the form $\rho(t) = \rho_0 \sin(\pi t/50)$. The numerical results obtained for neutron density using sinusoidal reactivity have been presented in Table 7.5. The obtained results justify that the HWOM is an accurate and efficient computational technique for solving point kinetic equation in the presence of nonlinear sinusoidal reactivity insertion. It can be observed that there is a good agreement between the obtained results and the other available method results [88]. More accurate results for large time can be obtained with increase in collocation points, namely, high level of resolution. Figure 7.1 represents the neutron density for sinusoidal reactivity obtained by the Haar wavelet method.

7.5.4.2 For Group of Six Delayed Neutron

To check the accuracy of the Haar wavelet method, it is applied to the thermal reactor with the following parameters from McMohan and Pierson [64] and Patra and Saha Ray [66]: $\lambda_1 = 0.0124\,s^{-1}$, $\lambda_2 = 0.0305\,s^{-1}$, $\lambda_3 = 0.0111\,s^{-1}$, $\lambda_4 = 0.301\,s^{-1}$, $\lambda_5 = 1.14\,s^{-1}$, $\lambda_6 = 3.01\,s^{-1}$, $\beta_1 = 0.000215$, $\beta_2 = 0.001424$,

TABLE 7.4

The Neutron Density $n(t)$ of the Thermal Reactors with Zig-Zag Reactivity

Time (s)	EPCA [89]	HWOM with $m = 256$
0.5	1.721422424E+00	1.721419616600E+00
1	1.211127414E+00	1.211121251790E+00
1.5	1.892226142E+00	1.892197202560E+00
2	2.521600530E+00	2.521593863891E+00
10	1.204710536E+01	1.204676703380E+01
100	6.815556889E+07	6.792200468046E+07

TABLE 7.5

The Neutron Density $n(t)$ of the Fast Reactor with Sinusoidal Reactivity

Time (s)	BEFD [88]	HWOM with $m = 256$
10	2.065383519E+00	2.065379972801E+00
20	8.854133921E+00	8.854133966731E+00
30	4.064354222E+01	4.064695450288E+01
40	6.135607517E+01	6.1364077105064E+01
50	4.610628770E+01	4.6114226208650E+01
60	2.912634840E+01	2.9133486699766E+01
70	1.895177042E+01	1.8958245424550E+01
80	1.393829211E+01	1.3944612597243E+01
90	1.253353406E+01	1.2540632819535E+01
100	1.544816514E+01	1.5480673682222E+01

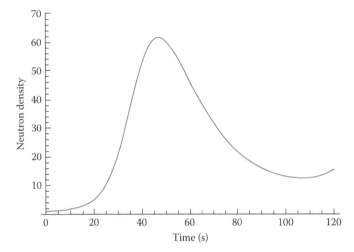

FIGURE 7.1
Neutron density for sinusoidal reactivity.

$\beta_3 = 0.001274$, $\beta_4 = 0.002568$, $\beta_5 = 0.000748$, $\beta_6 = 0.000273$, $\beta = 0.006502$, and $l = 0.0005$ s. The reactivity is a time-dependent function of the form $\rho(t) = \beta \sin(\pi t/5)$. The numerical results obtained for neutron density using sinusoidal reactivity have been presented in Table 7.6. The obtained results justify that the Haar wavelet method is an efficient computational technique for solving point kinetic equation using sinusoidal reactivity insertions. It can be observed that there is a good agreement between the obtained results and the other available method results [63,64]. More accurate results for large time can be obtained with increase in collocation points. Figure 7.2 illustrates the neutron density for sinusoidal reactivity obtained by the Haar wavelet method.

TABLE 7.6

Results Obtained for Neutron Density $n(t)$ with Sinusoidal
Reactivity Function

Time (s)	Taylor [64]	CORE [63]	HWOM with Collocation Points $m = 32$
2.0	11.3820	10.1475	11.3131
4.0	92.2761	96.7084	90.3934
6.0	16.9149	16.9149	15.6511
8.0	8.8964	8.8964	8.52265
10.0	13.1985	13.1985	13.1255

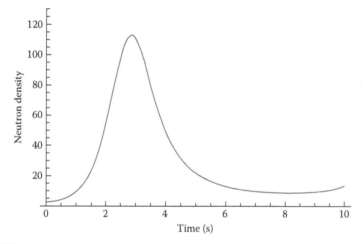

FIGURE 7.2
Neutron density for sinusoidal reactivity calculated with the Haar wavelet method.

7.5.5 Pulse Reactivity Insertion

In the present case, we consider the effect of pulse reactivity function [89]

$$\rho(t) = \begin{cases} 4\beta(e^{-2t^2}), & t < 1 \\ 0, & t > 1 \end{cases}$$

in nuclear reactor for one-group delayed neutron, that is, $M = 1$ and the parameters are used as follows from the works of Picca et al. [89]: $\lambda_1 = 0.077 \text{ s}^{-1}$, $\beta_1 = 0.006502 = \beta$, neutron source $q = 0$, and $l = 5 \times 10^{-4}$ s. The results for neutron density using pulse reactivity are provided in Table 7.7 and illustrated in Figure 7.3. It can be observed that there is a good agreement between the obtained results and the other available method results [89,98]. This curve has the expected shape for pulse reactivity. Here, we consider the number of collocation points $m = 32$.

TABLE 7.7

Results Obtained for Neutron Density $n(t)$ with Pulse Reactivity Function

Time (s)	EPCA [89]	CATS [98]	Present Haar Wavelet Method
0.5	9.38004427E+06	9.380044272E+06	9.383312098E+06
0.8	1.69477616E+08	1.694776161E+08	1.6963295571E+08
1	1.075131704E+08	1.075131704E+08	1.07551610768E+08
2	4.834117624E+06	4.834106369E+06	4.8410195087E+06
3	4.833903589E+06	4.833892339E+06	4.8516408749E+06

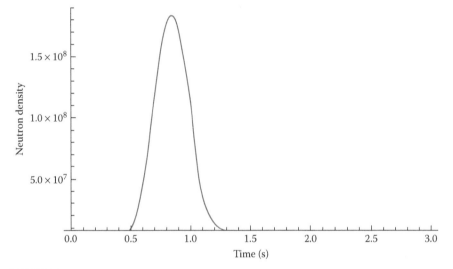

FIGURE 7.3
Neutron density calculated using the Haar wavelet method in response to pulse reactivity.

7.6 Convergence Analysis and Error Estimation

In this section, we have introduced the error analysis for the Haar wavelet method.

Lemma

Let, $f(x) \in L^2(R)$ be a continuous function defined in $(0,1]$. Then, the error norm at Jth-level resolution satisfies the following inequalities:

$$\|E_J\| \le \frac{\eta^2}{12} 2^{-2J}$$

where $|f'(x)| \le \eta$, $\forall x \in [0,1]$, and $\eta > 0$. Here, M is a positive number related to the Jth-level resolution of the wavelet given by $M = 2^J$.

Proof

$$|E_J| = |f(x) - f_J(x)| = \left| \sum_{i=2M}^{\infty} a_i h_i(x) \right|$$

where:

$$f_J(x) = \sum_{i=0}^{2M-1} a_i h_i(x), \quad \text{for } M = 2^j$$

$$\|E_J\|^2 = \int_{-\infty}^{\infty} \left[\sum_{i=2M}^{\infty} a_i h_i(x), \sum_{r=2M}^{\infty} a_r h_r(x) \right] dx$$

$$= \sum_{i=2M}^{\infty} \sum_{r=2M}^{\infty} a_i a_r \int_{-\infty}^{\infty} h_i(x) h_r(x) dx$$

$$\le \sum_{i=2M}^{\infty} |a_i|^2$$

Now, we consider $a_i = \int_0^1 2^{j/2} f(x) \varphi(2^j x - k) dx$

where $h_i(x) = 2^{j/2} h(2^j x - k)$, $k = 0,1,2,\ldots,2^j - 1$, and $j = 0,1,\ldots,J$.

$$h(2^j x - k) = \begin{cases} 1, & k2^{-j} \le x < \left(k + \frac{1}{2}\right)2^{-j} \\ -1, & \left(k + \frac{1}{2}\right)2^{-j} \le x < (k+1)2^{-j} \\ 0, & \text{elsewhere} \end{cases}$$

Therefore, applying mean value theorem

$$a_i = 2^{j/2} \left\{ \int_{k2^{-j}}^{[k+(1/2)]2^{-j}} f(x)dx - \int_{[k+(1/2)]2^{-j}}^{(k+1)2^{-j}} f(x)dx \right\}$$

$$= 2^{j/2} \left\{ \left[\left(k + \frac{1}{2}\right)2^{-j} - k2^{-j} \right] f(\psi_1) - \left[(k+1)2^{-j} - \left(k + \frac{1}{2}\right)2^{-j} \right] f(\psi_2) \right\}$$

where:

$$\psi_1 \in \left[k2^{-j}, \left(k + \frac{1}{2} \right) 2^{-j} \right]$$

$$\psi_2 \in \left[\left(k + \frac{1}{2} \right) 2^{-j}, (k+1)2^{-j} \right]$$

Consequently, $a_i = 2^{-(j/2)-1}(\psi_1 - \psi_2)[f'(\psi)]$, applying Lagrange's mean value theorem, where $\psi \in (\psi_1, \psi_2)$.
    This implies,

$$a_i^2 = 2^{-j-2}(\psi_2 - \psi_1)^2 f'(\psi)^2$$

$$\leq 2^{-j-2}2^{-2j}\eta^2, \quad \text{since } |f'(x)| \leq \eta$$

$$= 2^{-3j-2}\eta^2$$

Therefore,

$$\|E_J\|^2 \leq \sum_{i=2M}^{\infty} a_i^2 \leq \sum_{i=2M}^{\infty} 2^{-3j-2}\eta^2 \leq \eta^2 \sum_{j=J+1}^{\infty} \sum_{i=2}^{2^{j+1}-1} 2^{-3j-2}$$

$$= \eta^2 \sum_{j=J+1}^{\infty} 2^{-3j-2}\left(2^{j+1} - 1 - 2^j + 1\right) = \frac{\eta^2}{4} \sum_{j=J+1}^{\infty} 2^{-2j} \qquad (7.19)$$

$$= \frac{\eta^2}{4} \frac{2^{-2(J+1)}}{[1-(1/4)]} = \frac{\eta^2}{12} 2^{-2J}$$

From Equation 7.19, it can be observed that the error bound is inversely proportional to the level of resolution J. So, more accurate result can be obtained by increasing the level of resolution.

7.7 Conclusion

A numerical approximate solution has been presented in this chapter to explore the behavior of neutron density using reactivity functions. It can be seen that the proposed numerical method, the HWOM, certainly perform quite well. To predict the dynamical behavior for thermal and fast reactors with constant and time-dependent reactivity function and to obtain the solution of neutron point kinetic equation, the numerical method, namely,

HWOM, is undoubtedly very simple and efficient. This numerical technique provides a good result for the point reactor kinetic equations with a constant and time-dependent reactivity function. Results of this method have been compared with other available methods that exist in the open literature in order to show justification of the present method. The cited comparisons revealed that the obtained numerical solutions agree well with the other solutions in the open literature.

The accuracy of the obtained solutions are quite high even if the number of collocation points is small. By increasing the number of collocation points, the error of the approximation solution rapidly decreases. In a systematic comparison with other existing methods, it may be concluded that the present method is simple and efficient. This method is applied to different types of reactivity in order to check the validity of the proposed method. Moreover, the obtained approximate results have also been compared with other available numerical results that exist in the open literature. It manifests that the results obtained by the HWOM are in good agreement with other available results even for large time range and it is certainly simpler than other methods in the open literature. Haar wavelets are preferred due to their useful properties such as simple applicability, orthogonality, and compact support. The HWOM needs less computational effort as the major blocks of the HWOM are calculated only once and used in the subsequent computations repeatedly. Simply availability and fast convergence of the Haar wavelets provide a solid foundation for highly linear as well as nonlinear problems of differential equations. This proposed method with far less degrees of freedom and small computational overhead provides better solution. It can be concluded that this method is quite suitable, accurate, and efficient in comparison with other classical methods. The pertinent feature of the method is that the errors for solutions may be reduced for large value of m, namely, $m = 1024$ or more number of collocation points. The main advantage of this method is that it transfers the whole scheme into a system of algebraic equations, by virtue of it the computation is very easy and simple compared with other methods. This present analysis shows the applicability of the Haar wavelet method for the numerical solution of neutron point kinetic equation in nuclear reactor dynamics. The obtained results manifest the plausibility of the Haar wavelet method for neutron point kinetic equations.

8

Numerical Solution Using Two-Dimensional Haar Wavelet Collocation Method for Stationary Neutron Transport Equation in Homogeneous Isotropic Medium

8.1 Introduction

The term *neutron transport* denotes the study of the motions and interactions of neutrons with the atomic nuclei of the medium. The fractional neutron transport equation represents a linear case of the Boltzmann equation, and it has many applications in physics as well as in engineering. The neutron transport model in a nuclear reactor is an anomalous diffusion process. Anomalous diffusion is different from the normal diffusion and is characterized by features like slower or faster movement of diffusing particles. A useful characterization of the diffusion process is again through the scaling of the mean square displacement with time, which can be defined as $\langle x^2(t) \rangle \sim t^\gamma, \gamma \neq 1$. Diffusion is then classified through the scaling index γ. The case $\gamma = 1$ is a normal diffusion, and all other cases are termed anomalous. The cases $\gamma > 1$ form the family of superdiffusive processes, including the particular case $\gamma = 2$, which is called ballistic diffusion, and the cases $\gamma < 1$ are the subdiffusive processes. Hence, the solution of the fractional order transport model characterizes the dynamics of an anomalous process [99–101].

The motivation of this research work is to solve a typical problem of mathematical physics—the solving of a neutral particle transport equation that has numerous applications in physics [102,103]. In a nuclear reactor, the neutrons are generated by fission of the nucleus, and they are named as fast neutrons with an average speed equal to 2×10^7 m/s. In the thermal nuclear reactor, fast neutrons are subjected to a slowness process, decreasing their energy until they are in a state of equilibrium with the other atoms in the environment. The main goal in reactor theory is to find the neutron profile within the nuclear reactor core and hence the power distribution. The flux profile can be mathematically identified by the solution of an integrodifferential equation (IDE) known as neutron transport equation.

IDEs have many applications in different fields of mechanical [104] and nuclear engineering, chemistry, astronomy, biology, economics, potential theory, and electrostatics. An exact solution of this IDE has been found only in some specific cases. In many cases, determination of analytical solutions of IDEs is an unwieldy task; therefore, our aim is focused on determining an accurate and efficient numerical method [105].

In this study, we consider a linear form of the Boltzmann equation with a source function of the form $f(x,\eta) = A(\eta)\cos \pi\eta + B(\eta)\sin \pi\eta$. To obtain the solution of this stationary transport equation, we have applied the two-dimensional Haar wavelet transform method. Some numerical examples illustrate the advantage of this method applied to stationary transport equation.

Analysis of wavelet theory is a new branch of mathematics and widely applied in signal analysis, image processing, and numerical analysis [106,107]. Among the different wavelet families, mathematically most simple are the Haar wavelets [108]. In 1910, Alfred Haar introduced the notion of wavelets. Haar wavelets have the properties of orthogonality and normalization, having compact support and simple expression [94,95,97]. Due to their simplicity, Haar wavelets are an efficient and effective tool for solving both differential and integral equations.

8.2 Formulation of Neutron Transport Equation Model

The one-dimensional transport equation [99–101] can be presented as

$$\eta\frac{\partial\phi(x,\eta)}{\partial x} + \sigma_t\phi(x,\eta) = \int_{-1}^{1} \sigma_s(\eta,\eta')\phi(x,\eta')d\eta' + \frac{q(x)}{2} \tag{8.1}$$

with

$$\sigma_s(\eta,\eta') = \sum_{k=0}^{\infty} \frac{2k+1}{2}\sigma_{sk}P_k(\eta)P_k(\eta')$$

where:

σ_t is the total cross section

σ_{sl} with $l = 0,1,...,L$ are the components of differential scattering cross section

$P_k(\eta)$ is the Legendre polynomial of degree k

$q(x)$ is the source function

Here, we consider the IDE for the stationary case of transport theory [102] by considering $\sigma_t \equiv 1$, $\sigma_s(\eta,\eta') \equiv 1/2$, and $q(x) \equiv 2f(x,\eta)$ in Equation 8.1, which yields an IDE for the stationary case of transport theory [103]

$$\eta \frac{\partial \phi(x,\eta)}{\partial x} + \phi(x,\eta) = \frac{1}{2} \int_{-1}^{1} \phi(x,\eta') d\eta' + f(x,\eta),$$

(8.2)

$$\forall (x,\eta) \in D_1 \times D_2 = [0,1] \times [-1,1], D_2 = D_2' \cup D_2'' = [-1,0] \cup [0,1]$$

where:
  $\phi(x,\eta)$ is the neutron density that migrate in a direction that makes an
    angle μ with the x-axis and $\eta = \cos \mu$
  $f(x,\eta)$ is a given source function

The following are the boundary conditions:

$$\phi(0,\eta) = 0 \text{ if } \eta > 0 \text{ and } \phi(1,\eta) = 0 \text{ if } \eta < 0$$

(8.3)

Now, we split the Equation 8.2 into two equations using the following notations:

$$\phi^+ = \phi(x,\eta) \text{ if } \eta > 0 \text{ and } \phi^- = \phi(x,-\eta) \text{ if } \eta < 0$$

(8.4)

By denoting, $\eta' = -\eta$, we can obtain

$$\int_{-1}^{0} \phi(x,\eta') d\eta' = \int_{0}^{1} \phi(x,-\eta) d\eta = \int_{0}^{1} \phi^- d\eta$$

In view of Equation 8.4, Equation 8.2 can be written as

$$\eta \frac{\partial \phi^+}{\partial x} + \phi^+ = \frac{1}{2} \int_{0}^{1} (\phi^+ + \phi^-) d\eta' + f^+ \text{ for } \eta > 0$$

(8.5)

$$-\eta \frac{\partial \phi^-}{\partial x} + \phi^- = \frac{1}{2} \int_{0}^{1} (\phi^+ + \phi^-) d\eta' + f^- \text{ for } \eta < 0$$

(8.6)

with the boundary conditions $\phi^+(0,\eta) = 0$ and $\phi^-(1,\eta) = 0$.
  Adding and subtracting the Equations 8.5 and 8.6 and then introducing the following dependent variables, we obtain

$$u = \frac{1}{2}(\phi^+ + \phi^-), v = \frac{1}{2}(\phi^+ - \phi^-), g = \frac{1}{2}(f^+ + f^-), \text{ and } r = \frac{1}{2}(f^+ - f^-) \quad (8.7)$$

We also obtain the following system:

$$\eta \frac{\partial v}{\partial x} + u = \int_{0}^{1} u d\eta + g$$

(8.8)

$$\eta \frac{\partial u}{\partial x} + v = r \tag{8.9}$$

along with the following boundary conditions

$$\left. \begin{array}{lll} u+v=0, & \text{for} & x=0 \\ u-v=0, & \text{for} & x=1 \end{array} \right\} \tag{8.10}$$

Eliminating the value of v from Equations 8.8 and 8.9, we rewrite the Equations 8.8 through 8.10 in the following form:

$$-\eta^2 \frac{\partial^2 u}{\partial x^2} + u = \int_0^1 u d\eta + g - \eta \frac{\partial r}{\partial x} \tag{8.11}$$

$$\left(u - \eta \frac{\partial u}{\partial x} \right)\bigg|_{x=0} = -r(0,\eta) \tag{8.12}$$

$$\left(u + \eta \frac{\partial u}{\partial x} \right)\bigg|_{x=1} = r(1,\eta) \tag{8.13}$$

where:
$\eta \in [0,1]$

8.3 Mathematical Model of the Stationary Neutron Transport Equation in a Homogeneous Isotropic Medium

Let us consider the stationary transport equation:

$$\eta \frac{\partial \phi(x,\eta)}{\partial x} + \phi(x,\eta) = \frac{1}{2} \int_{-1}^1 \phi(x,\eta') d\eta' + f(x,\eta) \tag{8.14}$$

where:

$$f(x,\eta) = P(\eta)\cos \pi x + Q(\eta)\sin \pi x \tag{8.15}$$

with $P(\eta)$ be an odd function and $Q(\eta)$ be an even function.
   The equation is accompanied by boundary conditions:

$$\phi(0,\eta) = 0 \text{ if } \eta > 0 \text{ and } \phi(1,\eta) = 0 \text{ if } \eta < 0$$

According to the notations introduced in Equation 8.7, the functions g and r can be obtained as

$$g(x, \eta) = Q(\eta)\sin \pi x \text{ and } r(x, \eta) = P(\eta)\cos \pi x \qquad (8.16)$$

and u satisfies the equation

$$\eta^2 \frac{\partial^2 u}{\partial x^2} - u + \int_0^1 u(x, \eta')d\eta' + [Q(\eta) + \pi\eta P(\eta)]\sin \pi x = 0 \qquad (8.17)$$

with the boundary conditions

$$u(0, \eta) - \eta \frac{du(0, \eta)}{dx} = -r(0, \eta) = -P(\eta) \qquad (8.18)$$

$$u(1, \eta) + \eta \frac{du(1, \eta)}{dx} = r(1, \eta) = -P(\eta) \qquad (8.19)$$

In particular case, we consider the source function

$$f(x, \eta) = \pi\eta^3 \cos \pi x + \left(\eta^2 - \frac{1}{3}\right)\sin \pi x$$

where:

$$P(\eta) = \pi\eta^3$$

$$Q(\eta) = \eta^2 - \frac{1}{3}$$

According to Equation 8.16, the functions g and r will be

$$g(x, \eta) = \left(\eta^2 - \frac{1}{3}\right)\sin \pi x \text{ and } r(x, \eta) = \pi\eta^3 \cos \pi x$$

and u satisfies the following equation:

$$\eta^2 \frac{\partial^2 u}{\partial x^2} - u + \int_0^1 u(x, \eta')d\eta' + \left(\eta^2 - \frac{1}{3} + \pi^2\eta^4\right)\sin \pi x = 0 \qquad (8.20)$$

The boundary conditions now become

$$\left. \begin{aligned} u(0, \eta) - \eta \frac{du(0, \eta)}{dx} &= -r(0, \eta) = -\pi\eta^3 \\[2ex] u(1, \eta) + \eta \frac{du(1, \eta)}{dx} &= r(1, \eta) = -\pi\eta^3 \end{aligned} \right\} \qquad (8.21)$$

In this case the exact solution is $u_{exact} = \phi_{exact}(x, \eta) = \eta^2 \sin \pi x$ and $v_{exact} = 0$.

8.4 Application of the Two-Dimensional Haar Wavelet Collocation Method to Solve the Stationary Neutron Transport Equation

Let us divide the intervals D_1 and D_2'' into m equal subintervals each of width $\Delta_1 = 1/m$ and $\Delta_2 = 1/m$, respectively. We assume

$$\frac{\partial^2 u(x,\eta)}{\partial x^2} \equiv D^2 u(x,\eta) = \sum_{i=1}^{m} \sum_{j=1}^{m} c_{ij} h_i(x) h_j(\eta) \qquad (8.22)$$

Integrating Equation 8.22 two times from 0 to x, we can obtain

$$u(x,\eta) = \sum_{i=1}^{m} \sum_{j=1}^{m} c_{ij} Q h_i(x) h_j(\eta) + C_1 + x C_2 \qquad (8.23)$$

where $Q h_i(x)$ is given by in Equation 7.8 for $\alpha = 2$.
Substituting $x = 0$ in Equation 8.23, we can obtain

$$u(0,\eta) = C_1 \qquad (8.24)$$

Similarly at the boundary $x = 1$, we have

$$u(1,\eta) = \sum_{i=1}^{m} \sum_{j=1}^{m} c_{ij} Q h_i(x) \Big|_{x=1} h_j(\eta) + C_1 + C_2 \qquad (8.25)$$

Next, together with Equations 8.24 and 8.25, we have

$$C_2 = u(1,\eta) - u(0,\eta) - \sum_{i=1}^{m} \sum_{j=1}^{m} c_{ij} Q h_i(x) \Big|_{x=1} h_j(\eta) \qquad (8.26)$$

Now, by using Equations 8.24 and 8.26 in Equation 8.23, we can obtain the following approximate solution:

$$u(x,\eta) = \sum_{i=1}^{m} \sum_{j=1}^{m} c_{ij} Q h_i(x) h_j(\eta) + u(0,\eta)$$

$$+ x \left[u(1,\eta) - u(0,\eta) - \sum_{i=1}^{m} \sum_{j=1}^{m} c_{ij} Q h_i(x) \Big|_{x=1} h_j(\eta) \right] \qquad (8.27)$$

In view of the conditions represented by Equations 8.2 and 8.21, we assume the following boundary conditions:

$$u(0,\eta) = 0 \text{ and } u(1,\eta) = 0 \qquad (8.28)$$

Now by using Equations 8.28 into Equation 8.27, we have

$$u(x,\eta) = \sum_{i=1}^{m}\sum_{j=1}^{m} c_{ij}Qh_i(x)h_j(\eta) - x\left[\sum_{i=1}^{m}\sum_{j=1}^{m} c_{ij}Qh_i(x)\bigg|_{x=1} h_j(\eta)\right] \qquad (8.29)$$

Then by using Equation 8.22 and 8.29 into Equation 8.20, we obtain the numerical scheme as

$$\sum_{i=1}^{m}\sum_{j=1}^{m} c_{ij}h_i(x)h_j(\eta) - \left(\frac{1}{\eta^2}\right)\left\{\sum_{i=1}^{m}\sum_{j=1}^{m} c_{ij}Qh_i(x)h_j(\eta) - x\left[\sum_{i=1}^{m}\sum_{j=1}^{m} c_{ij}Qh_i(x)\bigg|_{x=1} h_j(\eta)\right]\right\}$$

$$+\left(\frac{1}{\eta^2}\right)\left\{\sum_{i=1}^{m}\sum_{j=1}^{m} c_{ij}Qh_i(x)\int_0^1 h_j(\eta')d\eta' - x\left[\sum_{i=1}^{m}\sum_{j=1}^{m} c_{ij}Qh_i(x)\bigg|_{x=1}\int_0^1 h_j(\eta')d\eta'\right]\right\} \qquad (8.30)$$

$$+\left(\frac{1}{\eta^2}\right)\left(\eta^2 - \frac{1}{3} + \pi^2\eta^4\right)\sin \pi x = 0$$

or we can write

$$\sum_{i=1}^{m}\sum_{j=1}^{m} c_{ij}h_i(x)h_j(\eta) - \left(\frac{1}{\eta^2}\right)\left\{\sum_{i=1}^{m}\sum_{j=1}^{m} c_{ij}Qh_i(x)h_j(\eta) - x\left[\sum_{i=1}^{m}\sum_{j=1}^{m} c_{ij}Qh_i(x)\bigg|_{x=1} h_j(\eta)\right]\right\}$$

$$+\left(\frac{1}{\eta^2}\right)\left\{\sum_{i=1}^{m}\sum_{j=1}^{m} c_{ij}Qh_i(x)\tilde{P}_j(\eta) - x\left[\sum_{i=1}^{m}\sum_{j=1}^{m} c_{ij}Qh_i(x)\bigg|_{x=1}\tilde{P}_j(\eta)\right]\right\} \qquad (8.31)$$

$$+\left(\frac{1}{\eta^2}\right)\left(\eta^2 - \frac{1}{3} + \pi^2\eta^4\right)\sin \pi x = 0$$

where:

$$\tilde{P}_j(\eta) = \int_0^1 h_j(\eta')d\eta' = \begin{cases} 1, & j=1 \\ 0, & j\neq 1 \end{cases}$$

By considering all the wavelet collocation points $\eta_l = A + (l-0.5)\Delta_2$ for $l = 1,2,\cdots,m$ and $x_k = A + (k-0.5)\Delta_1$ for $k = 1,2,\cdots,m$, we may obtain the system of m^2 number of algebraic equations involving m^2 unknown coefficients c_{ij}. By solving these system of equations using any mathematical software, we can obtain the Haar coefficients c_{ij}. Hence, we can obtain the approximate solution for neutron density $u(x,\eta)$ or $\phi(x,\eta)$ for stationary neutron transport equation.

8.5 Numerical Results and Discussions for Stationary Integer Order Neutron Transport Equation

Let us consider the numerical example for stationary neutron transport equation [103]

$$\eta^2 \frac{\partial^2 u}{\partial x^2} - u + \int_0^1 u(x, \eta')d\eta' + \left[Q(\eta)\right] + \pi\eta P(\eta)\sin \pi x = 0$$

where $P(\eta) = \pi\eta^3$ and $Q(\eta) = \eta^2 - 1/3$ with the following boundary conditions:

$$\left. \begin{aligned} u(0, \eta) - \eta \frac{du(0, \eta)}{dx} &= -r(0, \eta) = -\pi\eta^3 \\ u(1, \eta) + \eta \frac{du(1, \eta)}{dx} &= r(1, \eta) = -\pi\eta^3 \end{aligned} \right\}$$

We obtain the numerical approximate solutions for neutron density with $x \in [0,1]$ and with $\eta = 0.2, 0.4, 0.6,$ and 0.8. Here, we compare the numerical results with the exact solutions, and the absolute errors thus obtained are shown in Tables 8.1 through 8.4. Figure 8.1a through 8.1g illustrates the graphical comparison between the exact and numerical solutions for different values of x and η. Here, we have considered $m = 16$ and $m = 32$.

TABLE 8.1

Numerical Solution for Neutron Density at $\eta = 0.2$

			$\eta = 0.2$		
x	$u_{approximate}$ at $m = 16$	$u_{approximate}$ at $m = 32$	u_{exact}	Absolute Error for $m = 16$	Absolute Error for $m = 32$
0	0	0	0	0	0
0.1	0.0145110	0.0126787	0.0123607	0.00215028	0.000318068
0.2	0.0276062	0.0241160	0.0235114	0.00409481	0.000604573
0.3	0.0379918	0.0331925	0.0323607	0.00563116	0.000831859
0.4	0.0446648	0.0390202	0.0380423	0.00662251	0.000977905
0.5	0.0469623	0.0410283	0.04	0.00696235	0.001028320
0.6	0.0446648	0.0390202	0.0380423	0.00662251	0.000977905
0.7	0.0379918	0.0331925	0.0323607	0.00563116	0.000831859
0.8	0.0276062	0.0241160	0.0235114	0.00409481	0.000604573
0.9	0.0145110	0.0126787	0.0123607	0.00215028	0.000318068
1	-5.55112×10^{-16}	0	2.26622×10^{-17}	7.81733×10^{-17}	2.26622×10^{-17}

TABLE 8.2

Numerical Solution for Neutron Density at $\eta = 0.4$

	$\eta = 0.4$				
x	$u_{\text{approximate}}$ at $m = 16$	$u_{\text{approximate}}$ at $m = 32$	u_{exact}	Absolute Error for $m = 16$	Absolute Error for $m = 32$
0	0	0	0	0	0
0.1	0.050795	0.047102	0.0494427	0.00135228	0.00234008
0.2	0.096634	0.089593	0.0940456	0.00258878	0.00445269
0.3	0.132989	0.123313	0.1294430	0.00354611	0.00612958
0.4	0.156347	0.144963	0.1521690	0.00417811	0.00720577
0.5	0.164390	0.152424	0.16	0.00438973	0.00757626
0.6	0.156347	0.144963	0.1521690	0.00417811	0.00720577
0.7	0.132989	0.123313	0.1294430	0.00354611	0.00612958
0.8	0.096634	0.089593	0.0940456	0.00258878	0.00445269
0.9	0.050795	0.047102	0.0494427	0.00135228	0.00234008
1	-1.1103×10^{-16}	1.11022×10^{-16}	9.06486×10^{-17}	2.01671×10^{-17}	2.03737×10^{-17}

TABLE 8.3

Numerical Solution for Neutron Density at $\eta = 0.6$

	$\eta = 0.6$				
x	$u_{\text{approximate}}$ at $m = 16$	$u_{\text{approximate}}$ at $m = 32$	u_{exact}	Absolute Error for $m = 16$	Absolute Error for $m = 32$
0	0	0	0	0	0
0.1	0.108707	0.114695	0.111246	0.00253901	0.00344893
0.2	0.206809	0.218159	0.211603	0.00479398	0.00655639
0.3	0.284611	0.300268	0.291246	0.00663480	0.00902171
0.4	0.334601	0.352986	0.34238	0.00777956	0.01060560
0.5	0.351813	0.371152	0.36	0.00818717	0.01115220
0.6	0.334601	0.352986	0.34238	0.00777956	0.01060560
0.7	0.284611	0.300268	0.291246	0.00663480	0.00902171
0.8	0.206809	0.218159	0.211603	0.00479398	0.00655639
0.9	0.108707	0.114695	0.111246	0.00253901	0.00344893
1	4.44089×10^{-16}	0	2.03959×10^{-16}	2.4013×10^{-16}	2.03959×10^{-16}

8.6 Mathematical Model for Fractional Order Stationary Neutron Transport Equation

The fractional order neutron transport is the process of anomalous diffusion. This model removes the lacunae of the conventional integer order model of neutron movements. This research analysis based on the Haar wavelet is

TABLE 8.4

Numerical Solution for Neutron Density at $\eta = 0.8$

		$\eta = 0.8$	
x	$u_{approximate}$ at $m = 32$	u_{exact}	Absolute Error for $m = 32$
0	0	0	0
0.1	0.188276	0.197771	0.0094945
0.2	0.358184	0.376183	0.0179982
0.3	0.492935	0.517771	0.0248355
0.4	0.579515	0.608676	0.0291610
0.5	0.609326	0.64	0.0306743
0.6	0.579515	0.608676	0.0291610
0.7	0.492935	0.517771	0.0248355
0.8	0.358184	0.376183	0.0179982
0.9	0.188276	0.197771	0.0094953
1	4.44089×10^{-16}	3.62594×10^{-16}	8.14947×10^{-17}

probably being performed for the first time for the fractional order model of steady-state neutron transport. The analysis carried out in this present study thus forms a crucial step in the process of development of fractional order transport model for a nuclear reactor.

In view of Equation 8.11, let us consider the fractional order ($\beta > 0$) stationary neutron transport equation

$$\eta^2 \frac{\partial^\beta u}{\partial x^\beta} - u + \int_0^1 u(x, \eta')d\eta' + g - \eta \frac{\partial r}{\partial x} = 0, \quad 1 < \beta \le 2 \qquad (8.32)$$

with the boundary conditions

$$\left(u - \eta \frac{\partial u}{\partial x} \right)\bigg|_{x=0} = -r(0, \eta) \qquad (8.33)$$

$$\left(u + \eta \frac{\partial u}{\partial x} \right)\bigg|_{x=1} = r(1, \eta) \qquad (8.34)$$

where:
$\eta \in [0, 1]$

- *Case I*: In this case, we consider the source function

$$f(x, \eta) = \pi \eta^3 \cos \pi x + \left(\eta^2 - \frac{1}{3} \right) \sin \pi x \qquad (8.35)$$

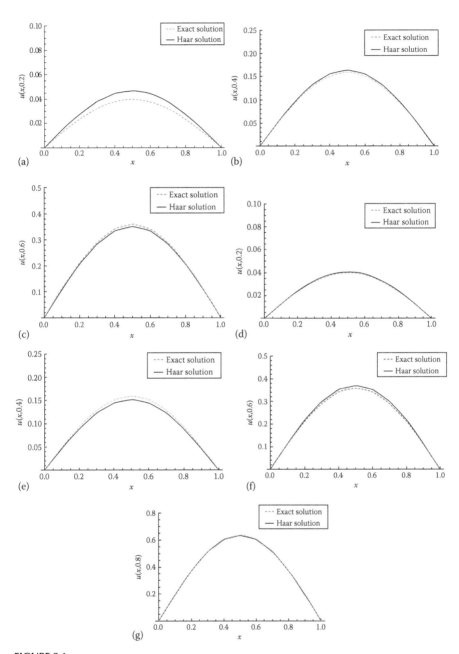

FIGURE 8.1
Comparison between the exact and numerical approximate solutions for neutron density at
(a) $\eta = 0.2$ and $m = 16$; (b) $\eta = 0.4$ and $m = 16$; (c) $\eta = 0.6$ and $m = 16$; (d) $\eta = 0.2$ and $m = 32$; (e) $\eta = 0.4$
and $m = 32$; (f) $\eta = 0.4$ and $m = 32$; (g) $\eta = 0.6$ and $m = 32$.

According to Equation 8.7, the functions g and r can be defined as

$$g(x,\eta) = \left(\eta^2 - \frac{1}{3}\right)\sin \pi x \text{ and } r(x,\eta) = \pi\eta^3 \cos \pi x$$

Therefore, u satisfies the equation

$$\eta^2 \frac{\partial^\beta u}{\partial x^\beta} - u + \int_0^1 u(x,\eta')d\eta' + \left(\eta^2 - \frac{1}{3} + \pi^2\eta^4\right)\sin \pi x = 0 \qquad (8.36)$$

The boundary conditions now become

$$
\left.
\begin{aligned}
u(0,\eta) - \eta\frac{du(0,\eta)}{dx} &= -r(0,\eta) = -\pi\eta^3 \\[2mm]
u(1,\eta) + \eta\frac{du(1,\eta)}{dx} &= r(1,\eta) = -\pi\eta^3
\end{aligned}
\right\} \qquad (8.37)
$$

Here, the integer order, that is, classical solution [103] is

$$u_{\text{classical}} = \phi_{\text{classical}}(x,\eta) = \eta^2 \sin \pi x \qquad (8.38)$$

- *Case II*: In this case, we consider the source function as

$$f(x,\eta) = -2\pi^2\eta^4 \cos(2\pi x) + \left(\eta^2 - \frac{2}{3}\right)\sin^2 \pi x \qquad (8.39)$$

Therefore, according to Equation 8.7, the functions g and r will be

$$g(x,\eta) = f(x,\eta) = -2\pi^2\eta^4 \cos(2\pi x) + \left(\eta^2 - \frac{2}{3}\right)\sin^2 \pi x \text{ and } r(x,\eta) = 0$$

Therefore, u satisfies the equation

$$\eta^2 \frac{\partial^\beta u}{\partial x^\beta} - u + \int_0^1 u(x,\eta')d\eta' + \left[-2\pi^2\eta^4 \cos(2\pi x) + \left(\eta^2 - \frac{2}{3}\right)\sin^2 \pi x\right] = 0 \quad (8.40)$$

The boundary conditions now become

$$
\left.
\begin{aligned}
u(0,\eta) - \eta\frac{du(0,\eta)}{dx} &= 0 \\[2mm]
u(1,\eta) + \eta\frac{du(1,\eta)}{dx} &= 0
\end{aligned}
\right\} \qquad (8.41)
$$

In this case, the integer order, that is, the classical solution is as follows [109]:

$$u_{\text{classical}} = \phi_{\text{classical}}(x,\eta) = \eta^2 \sin^2 \pi x \qquad (8.42)$$

8.7 Application of the Two-Dimensional Haar Wavelet Collocation Method to the Fractional Order Stationary Neutron Transport Equation

Let us divide the intervals D_1 and D_2'' into m equal subintervals each of width $\Delta_1 = 1/m$ and $\Delta_2 = 1/m$, respectively. We assume

$$\frac{\partial^\beta u(x,\eta)}{\partial x^\beta} \equiv D^\beta u(x,\eta) = \sum_{i=1}^{m}\sum_{j=1}^{m} c_{ij} h_i(x) h_j(\eta) \tag{8.43}$$

Operating Riemann–Liouville fractional integral operator J^β on both sides of Equation 8.43, we obtain

$$u(x,\eta) = \sum_{i=1}^{m}\sum_{j=1}^{m} c_{ij} Q^\beta h_i(x) h_j(\eta) + C_1 + x C_2 \tag{8.44}$$

where $Q^\beta h_i(x)$ is given by Equation 7.8 for $\alpha = \beta$.
Substituting $x = 0$ into Equation 8.44, we can obtain

$$u(0,\eta) = C_1 \tag{8.45}$$

Similarly, at the boundary $x = 1$, we have

$$u(1,\eta) = \sum_{i=1}^{m}\sum_{j=1}^{m} c_{ij} Q^\beta h_i(x)\Big|_{x=1} h_j(\eta) + C_1 + C_2 \tag{8.46}$$

Next, together with Equations 8.45 and 8.46, we have

$$C_2 = u(1,\eta) - u(0,\eta) - \sum_{i=1}^{m}\sum_{j=1}^{m} c_{ij} Q^\beta h_i(x)\Big|_{x=1} h_j(\eta) \tag{8.47}$$

Now, by using Equations 8.45 and 8.47 in Equation 8.44, we can obtain the following approximate solution:

$$u(x,\eta) = \sum_{i=1}^{m}\sum_{j=1}^{m} c_{ij} Q^\beta h_i(x) h_j(\eta) + u(0,\eta)$$

$$\tag{8.48}$$

$$+ x\left[u(1,\eta) - u(0,\eta) - \sum_{i=1}^{m}\sum_{j=1}^{m} c_{ij} Q^\beta h_i(x)\Big|_{x=1} h_j(\eta) \right]$$

In view of conditions (Equations 8.2 and 8.41), we assume the following boundary conditions:

$$u(0,\eta) = u(1,\eta) = 0 \tag{8.49}$$

Now by using the above equation in Equation 8.48, we have

$$u(x,\eta) = \sum_{i=1}^{m}\sum_{j=1}^{m} c_{ij}Q^{\beta}h_i(x)h_j(\eta) - x\left[\sum_{i=1}^{m}\sum_{j=1}^{m} c_{ij}Q^{\beta}h_i(x)\Big|_{x=1} \; h_j(\eta)\right] \tag{8.50}$$

Then by using Equations 8.43 and 8.50 into Equations 8.36 and 8.40, we obtain the numerical scheme for case I as

$$\sum_{i=1}^{m}\sum_{j=1}^{m} c_{ij}h_i(x)h_j(\eta) - \left(\frac{1}{\eta^2}\right)\left\{\sum_{i=1}^{m}\sum_{j=1}^{m} c_{ij}Q^{\beta}h_i(x)h_j(\eta) - x\left[\sum_{i=1}^{m}\sum_{j=1}^{m} c_{ij}Q^{\beta}h_i(x)\Big|_{x=1}\; h_j(\eta)\right]\right\}$$

$$+ \left(\frac{1}{\eta^2}\right)\left\{\sum_{i=1}^{m}\sum_{j=1}^{m} c_{ij}Q^{\beta}h_i(x)\int_0^1 h_j(\eta')d\eta' - x\left[\sum_{i=1}^{m}\sum_{j=1}^{m} c_{ij}Q^{\beta}h_i(x)\Big|_{x=1}\int_0^1 h_j(\eta')d\eta'\right]\right\} \tag{8.51}$$

$$+ \left(\frac{1}{\eta^2}\right)\left(\eta^2 - \frac{1}{3} + \pi^2\eta^4\right)\sin\pi x = 0$$

and we also obtain the numerical scheme for case II as

$$\sum_{i=1}^{m}\sum_{j=1}^{m} c_{ij}h_i(x)h_j(\eta) - \left(\frac{1}{\eta^2}\right)\left\{\sum_{i=1}^{m}\sum_{j=1}^{m} c_{ij}Q^{\beta}h_i(x)h_j(\eta) - x\left[\sum_{i=1}^{m}\sum_{j=1}^{m} c_{ij}Q^{\beta}h_i(x)\Big|_{x=1}\; h_j(\eta)\right]\right\}$$

$$+ \left(\frac{1}{\eta^2}\right)\left\{\sum_{i=1}^{m}\sum_{j=1}^{m} c_{ij}Q^{\beta}h_i(x)\tilde{P}_j(\eta) - x\left[\sum_{i=1}^{m}\sum_{j=1}^{m} c_{ij}Q^{\beta}h_i(x)\Big|_{x=1}\; \tilde{P}_j(\eta)\right]\right\} \tag{8.52}$$

$$+ \left(\frac{1}{\eta^2}\right)\left[-2\pi^2\eta^4\cos(2\pi x) + \left(\eta^2 - \frac{2}{3}\right)\sin^2\pi x\right] = 0$$

where:

$$\tilde{P}_j(\eta) = \int_0^1 h_j(\eta')d\eta' = \begin{cases} 1, & j=1 \\ 0, & j\neq 1 \end{cases}$$

By considering all the wavelet collocation points $\eta_l = A + (l - 0.5)\Delta_2$ for $l = 1, 2, \cdots, m$ and $x_k = A + (k - 0.5)\Delta_1$ for $k = 1, 2, \cdots, m$, we may obtain the system of m^2 number of algebraic equations involving m^2 unknown coefficients c_{ij}. By solving this system of equations, we obtain the Haar coefficients c_{ij}. Hence we can obtain an approximate solution for neutron density $u(x, \eta)$ or $\phi(x, \eta)$ for fractional order stationary neutron transport equation.

8.8 Numerical Results and Discussions for Fractional Order Neutron Transport Equation

In the present section, we have considered two test problems for the solution of fractional order stationary neutron transport equation [103,109]. In cases I and II, we obtain approximate numerical solutions for neutron density for $x \in [0, 1]$ with $\eta = 0.2, 0.4, 0.6,$ and 0.8 and $\eta = 0.3, 0.5, 0.7,$ and 0.9, respectively. Here, we compare the numerical results with the exact or integer order classical solutions by considering fractional order $\beta = 1.94, 1.96, 1.98,$ and 2. For case I, the numerical solution obtained by the Haar wavelet collocation method (HWCM) have been displayed in Tables 8.5 through 8.8 and for case II, the numerical solution obtained by HWCM has been displayed in Tables 8.9 through 8.12. Here, in both cases we have taken $m = 16$. Figures 8.2a through 8.2d and 8.3a through 8.3d show the graphical comparison between

TABLE 8.5

Numerical Solution of Neutron Density When $\eta = 0.2$ for Case I

| | $\eta = 0.2$ and $m = 16$ | | | | |
| | Approximate Solution of u for Fractional Order | | | | $\phi_{classical} = u_{classical}$ Given in |
x	$\beta = 1.94$	$\beta = 1.96$	$\beta = 1.98$	$\beta = 2$	Equation 8.38
0	0	0	0	0	0
0.1	0.018967	0.017470	0.015984	0.014511	0.012360
0.2	0.035522	0.032889	0.030248	0.027606	0.023511
0.3	0.047793	0.044580	0.041309	0.037991	0.032360
0.4	0.054643	0.051439	0.048108	0.044664	0.038042
0.5	0.055636	0.052936	0.050039	0.046962	0.041031
0.6	0.051071	0.049175	0.047033	0.044664	0.038042
0.7	0.041835	0.040804	0.039517	0.037991	0.032360
0.8	0.029273	0.028927	0.028367	0.027606	0.023511
0.9	0.014864	0.014867	0.014748	0.014511	0.012360
1	1.665E–16	5.551E–17	2.775E–17	2.775E–17	2.266E–17

TABLE 8.6

Numerical Solution of Neutron Density When $\eta = 0.4$ for Case I

	$\eta = 0.4$ and $m = 16$				$\phi_{classical} = u_{classical}$
	Approximate Solution of u for Fractional Order				Given in
x	$\beta = 1.94$	$\beta = 1.96$	$\beta = 1.98$	$\beta = 2$	Equation 8.38
0	0	0	0	0	0
0.1	0.055022	0.053607	0.052197	0.050795	0.049442
0.2	0.104035	0.101578	0.099108	0.096634	0.094045
0.3	0.141999	0.139043	0.136036	0.132989	0.129443
0.4	0.165340	0.162439	0.159437	0.156347	0.152169
0.5	0.171963	0.169584	0.167054	0.064390	0.161100
0.6	0.161591	0.160023	0.158270	0.156347	0.152169
0.7	0.135634	0.134940	0.134054	0.132989	0.129443
0.8	0.097132	0.097127	0.096957	0.096634	0.094045
0.9	0.050239	0.050521	0.050704	0.050795	0.049442
1	9.064E–17	3.330E–16	0	0	9.0648E–17

TABLE 8.7

Numerical Solution of Neutron Density When $\eta = 0.6$ for Case I

	$\eta = 0.6$ and $m = 16$				$\phi_{classical} = u_{classical}$
	Approximate Solution of u for Fractional Order				Given in
x	$\beta = 1.94$	$\beta = 1.96$	$\beta = 1.98$	$\beta = 2$	Equation 8.38
0	0	0	0	0	0
0.1	0.115270	0.113072	0.110883	0.108707	0.111246
0.2	0.218149	0.214376	0.210594	0.206809	0.211603
0.3	0.298169	0.293702	0.289180	0.284611	0.291246
0.4	0.347777	0.343499	0.339103	0.334601	0.342380
0.5	0.362414	0.359056	0.355517	0.351813	0.360102
0.6	0.341260	0.339261	0.337036	0.334601	0.342380
0.7	0.287031	0.286459	0.285648	0.284611	0.291246
0.8	0.205932	0.206430	0.206718	0.206809	0.211603
0.9	0.106660	0.107470	0.108151	0.108707	0.111246
1	2.224E–16	0	0	4.440E–16	2.266E–17

the classical and numerical approximate solutions for the two test problems, respectively. In both the test problems, from Tables 8.5 through 8.12 it can be shown that the value of angular flux of neutrons ϕ increases with increase in the value of η. In practical applications, it should be concluded that the value of the density is equal to zero when the direction of the movement

TABLE 8.8

Numerical Solution of Neutron Density When $\eta = 0.8$ for Case I

	$\eta = 0.8$ and $m = 16$				$\phi_{classical} = u_{classical}$ Given in Equation 8.38
	Approximate Solution of u for Fractional Order				
x	$\beta = 1.94$	$\beta = 1.96$	$\beta = 1.98$	$\beta = 2$	
0	0	0	0	0	0
0.1	0.198969	0.195385	0.191820	0.188276	0.197771
0.2	0.376595	0.370464	0.364325	0.358184	0.376183
0.3	0.514830	0.507608	0.500306	0.492935	0.517771
0.4	0.600624	0.593758	0.586716	0.579515	0.608676
0.5	0.626066	0.620751	0.615164	0.609326	0.640001
0.6	0.589683	0.586629	0.588323	0.579515	0.608676
0.7	0.496110	0.495410	0.494346	0.492935	0.517771
0.8	0.356021	0.357059	0.357776	0.358184	0.376183
0.9	0.184429	0.185910	0.187190	0.188276	0.197771
1	0	2.220E–16	2.220E–16	0	3.623E–16

TABLE 8.9

Numerical Solution of Neutron Density When $\eta = 0.3$ for Case II

	$\eta = 0.3$ and $m = 16$				$\phi_{classical} = u_{classical}$ Given in Equation 8.42
	Approximate Solution of u for Fractional Order				
x	$\beta = 1.94$	$\beta = 1.96$	$\beta = 1.98$	$\beta = 2$	
0	0	0	0	0	0
0.1	0.015777	0.012747	0.009854	0.007094	0.008594
0.2	0.044167	0.038002	0.032073	0.026378	0.031094
0.3	0.074620	0.066278	0.058163	0.050283	0.058905
0.4	0.096080	0.087184	0.078388	0.069717	0.081405
0.5	0.100511	0.092802	0.084993	0.077123	0.090012
0.6	0.086444	0.081151	0.075559	0.069717	0.081405
0.7	0.059151	0.056593	0.053621	0.050283	0.058905
0.8	0.028955	0.028475	0.027605	0.026378	0.031094
0.9	0.006682	0.007048	0.007180	0.007094	0.008594
1	9.714E–17	6.938E–18	6.938E–18	3.035E–17	2.888E–19

of particles makes an angle of 90° to the x-axis. Figures 8.2a through 8.2d and 8.3a through 8.3d have shown that the solution has vanished at the physical domain. Figures 8.4a, 8.4b, 8.5a, and 8.5b display the convergence plot of absolute error for increasing collocation points $m = 16$ and 32 in cases I and II, respectively.

TABLE 8.10

Numerical Solution of Neutron Density When $\eta = 0.5$ for Case-II

	$\eta = 0.5$ and $m = 16$				$\phi_{classical} = u_{classical}$
	Approximate Solution of u for Fractional Order				Given in
x	$\beta = 1.94$	$\beta = 1.96$	$\beta = 1.98$	$\beta = 2$	Equation 8.42
0	0	0	0	0	0
0.1	0.035669	0.032509	0.029496	0.026624	0.023872
0.2	0.118640	0.111055	0.103752	0.096716	0.086372
0.3	0.215028	0.204108	0.193483	0.183154	0.163627
0.4	0.288103	0.276323	0.264715	0.253296	0.226127
0.5	0.309460	0.299676	0.289851	0.280010	0.250420
0.6	0.271311	0.265558	0.259541	0.253296	0.226127
0.7	0.188134	0.186868	0.185196	0.183154	0.163627
0.8	0.092364	0.094211	0.095653	0.096716	0.086372
0.9	0.020262	0.022630	0.024746	0.026624	0.023872
1	1.942E–16	1.179E–16	9.194E–17	5.052E–17	0

TABLE 8.11

Numerical Solution of Neutron Density When $\eta = 0.7$ for Case II

	$\eta = 0.7$ and $m = 16$				$\phi_{classical} = u_{classical}$
	Approximate Solution of u for Fractional Order				Given in
x	$\beta = 1.94$	$\beta = 1.96$	$\beta = 1.98$	$\beta = 2$	Equation 8.42
0	0	0	0	0	0
0.1	0.062774	0.057931	0.053321	0.048928	0.046790
0.2	0.212450	0.200322	0.188649	0.177420	0.169291
0.3	0.387385	0.369697	0.352498	0.335786	0.320709
0.4	0.520599	0.501516	0.482740	0.464282	0.443209
0.5	0.560127	0.544511	0.528864	0.513218	0.492200
0.6	0.491478	0.482783	0.473703	0.464282	0.443209
0.7	0.340763	0.339704	0.338031	0.335786	0.320709
0.8	0.166944	0.171045	0.174526	0.177420	0.169291
0.9	0.036162	0.040802	0.045052	0.048928	0.046790
1	2.567E–16	2.046E–16	1.249E–16	1.110E–16	0

TABLE 8.12

Numerical Solution of Neutron Density when $\eta = 0.9$ for Case II

	$\eta = 0.9$ and $m = 16$				$\varphi_{classical} = u_{classical}$ Given in Equation 8.42
	Approximate Solution of u for Fractional Order				
x	$\beta = 1.94$	$\beta = 1.96$	$\beta = 1.98$	$\beta = 2$	
0	0	0	0	0	0
0.1	0.099190	0.091723	0.084614	0.077846	0.077348
0.2	0.336621	0.317767	0.299627	0.282178	0.279848
0.3	0.614366	0.586809	0.560021	0.533993	0.530152
0.4	0.825999	0.796276	0.767040	0.738306	0.732652
0.5	0.888904	0.864663	0.840385	0.816115	0.810001
0.6	0.780004	0.766671	0.752752	0.738306	0.732652
0.7	0.540735	0.539414	0.537147	0.533993	0.530152
0.8	0.264761	0.271508	0.277298	0.282178	0.279848
0.9	0.057183	0.064668	0.071547	0.077846	0.077348
1	3.198E–16	2.081E–16	2.775E–17	3.156E–16	0

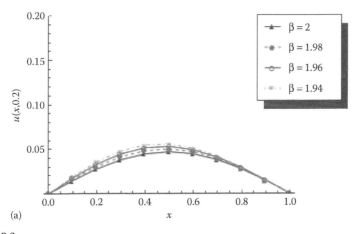

(a)

FIGURE 8.2
Comparison between the classical and fractional order numerical Haar solutions with $m = 16$ for neutron density in case I: (a) $\eta = 0.2$. *(Continued)*

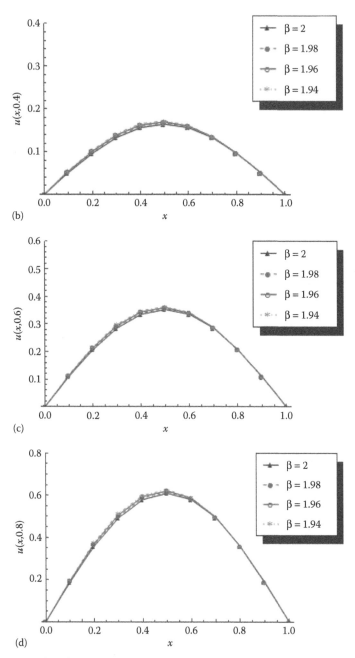

FIGURE 8.2 (Continued)
Comparison between the classical and fractional order numerical Haar solutions with $m = 16$ for neutron density in case I: (b) $\eta = 0.4$, (c) $\eta = 0.6$, (d) $\eta = 0.8$.

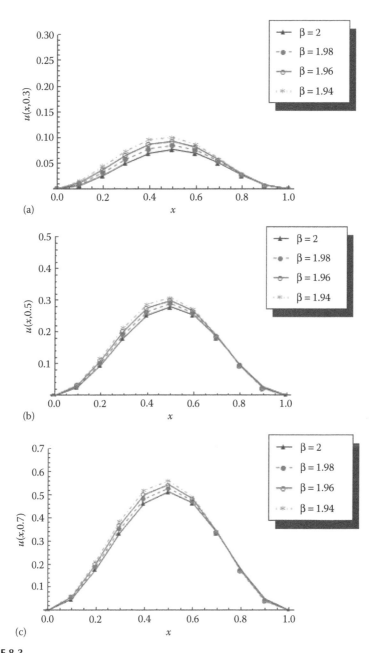

FIGURE 8.3
Comparison between the classical and fractional order numerical Haar solutions with $m = 16$ for neutron density in case II: (a) $\eta = 0.3$, (b) $\eta = 0.5$, (c) $\eta = 0.7$. *(Continued)*

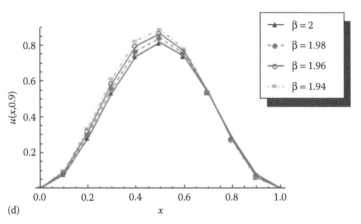

FIGURE 8.3 (Continued)
Comparison between the classical and fractional order numerical Haar solutions with $m = 16$ for neutron density in case II: (d) $\eta = 0.9$.

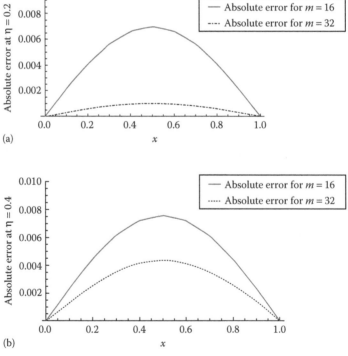

FIGURE 8.4
Absolute error for neutron density in case I at (a) $\eta = 0.2$, (b) $\eta = 0.4$.

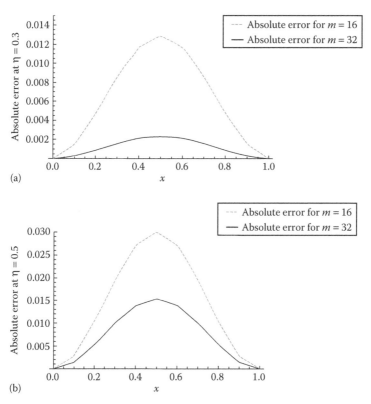

FIGURE 8.5
Absolute error for neutron density in case II at (a) $\eta = 0.3$, (b) $\eta = 0.5$.

8.9 Convergence Analysis of the Two-Dimensional Haar Wavelet Method

In this section, we have introduced the error analysis for the two-dimensional Haar wavelet method.

We assume that $f(x,y) \in C^2([a,b] \times [a,b])$ and there exist $M > 0$, for which

$$\left| \frac{\partial^2 f(x,y)}{\partial x \partial y} \right| \leq M, \forall x, y \in [a,b] \times [a,b]$$

Next, we may proceed as follows; suppose

$$f_{nm}(x,y) = \sum_{i=0}^{n-1} \sum_{j=0}^{m-1} c_{ij} h_i(x) h_j(y)$$

where:

$n = 2^{\alpha+1}, \alpha = 0,1,2,\ldots$

$m = 2^{\beta+1}$ and $\beta = 0,1,2,\ldots$

Then,

$$f(x,y) - f_{nm}(x,y) = \sum_{i=n}^{\infty}\sum_{j=m}^{\infty} c_{ij}h_i(x)h_j(y) + \sum_{i=n}^{\infty}\sum_{j=0}^{m-1} c_{ij}h_i(x)h_j(y) + \sum_{i=0}^{n-1}\sum_{j=m}^{\infty} c_{ij}h_i(x)h_j(y)$$

From Perseval's formula, we have

$$\left\| f(x,y) - f_{nm}(x,y) \right\|^2 = \int_a^b\int_a^b \left[f(x,y) - f_{nm}(x,y) \right]^2 dxdy$$

$$= \sum_{p=n}^{\infty}\sum_{s=m}^{\infty}\sum_{i=n}^{\infty}\sum_{j=m}^{\infty} c'_{ij}c'_{ps} \int_a^b h_i(x)h_p(x)dx \int_a^b h_j(y)h_s(y)dy$$

$$+ \sum_{p=n}^{\infty}\sum_{s=0}^{m-1}\sum_{i=n}^{\infty}\sum_{j=0}^{m-1} c'_{ij}c'_{ps} \int_a^b h_i(x)h_p(x)dx \int_a^b h_j(y)h_s(y)dy$$

$$+ \sum_{p=0}^{n-1}\sum_{s=m}^{\infty}\sum_{i=0}^{n-1}\sum_{j=m}^{\infty} c'_{ij}c'_{ps} \int_a^b h_i(x)h_p(x)dx \int_a^b h_j(y)h_s(y)dy$$

$$= \sum_{i=n}^{\infty}\sum_{j=m}^{\infty} {c'_{ij}}^2 + \sum_{i=n}^{\infty}\sum_{j=0}^{m-1} {c'_{ij}}^2 + \sum_{i=0}^{n-1}\sum_{j=m}^{\infty} {c'_{ij}}^2$$

where:

$$ {c'_{ij}}^2 = \frac{c_{ij}(b-a)^2}{2^{i+j}} $$

$$c_{ij} = \int_a^b \left[\int_a^b f(x,y)h_i(y)dy \right] h_j(x)dx$$

$$= \int_a^b \left[\int_{a+k\left[(b-a)/2^i\right]}^{a+\left[k+(1/2)\right]\left[(b-a)/2^i\right]} f(x,y)dy - \int_{a+\left[k+(1/2)\right]\left[(b-a)/2^i\right]}^{a+(k+1)\left[(b-a)/2^i\right]} f(x,y)dy \right] h_j(x)dx$$

Using the mean value theorem of integral calculus, we have

$$a+k\frac{(b-a)}{2^i}\le y_1\le a+\left(k+\frac{1}{2}\right)\frac{(b-a)}{2^i}, \quad a+\left(k+\frac{1}{2}\right)\frac{(b-a)}{2^i}\le y_2\le a+(k+1)\frac{(b-a)}{2^i}$$

Hence, we obtain

$$c_{ij}=(b-a)\int_a^b\left[f(x,y_1)2^{-i-1}-f(x,y_2)2^{-k-1}\right]h_j(x)dx$$

Again by using the mean value theorem,

$$c_{ij}=2^{-i-1}(b-a)\int_a^b\left[f(x,y_1)-f(x,y_2)\right]h_j(x)dx$$

Using Lagrange's mean value theorem,

$$c_{ij}=2^{-i-1}(b-a)\int_a^b\left[(y_1-y_2)\frac{\partial f(x,y^*)}{\partial y}\right]h_j(x)dx, \text{ where } y_1\le y^*\le y_2$$

$$=2^{-i-1}(b-a)(y_1-y_2)\left[\int_{a+k\left[(b-a)/2^j\right]}^{a+\left[k+(1/2)\right]\left[(b-a)/2^j\right]}\frac{\partial f(x,y^*)}{\partial y}dx-\int_{a+\left[k+(1/2)\right]\left[(b-a)/2^j\right]}^{a+(k+1)\left[(b-a)/2^j\right]}\frac{\partial f(x,y^*)}{\partial y}dx\right]$$

$$=2^{-i-1}(b-a)(y_1-y_2)\left[2^{-j-1}(b-a)\frac{\partial f}{\partial y}(x_1,y^*)-2^{-j-1}(b-a)\frac{\partial f}{\partial y}(x_2,y^*)\right]$$

Now, we use the mean value theorem of integral calculus:

$$a+k\frac{(b-a)}{2^j}\le x_1\le a+\left(k+\frac{1}{2}\right)\frac{(b-a)}{2^j}, \quad a+\left(k+\frac{1}{2}\right)\frac{(b-a)}{2^j}\le x_2\le a+(k+1)\frac{(b-a)}{2^j}$$

$$\le2^{-i-j-2}(b-a)^2(y_1-y_2)(x_1-x_2)\frac{\partial^2 f(x^*,y^*)}{\partial x\partial y}$$

But for $x_1 \leq x^* \leq x_2$, $(y_1 - y_2) \leq (b-a)$, and $(x_1 - x_2) \leq (b-a)$, we obtain

$$c_{ij} \leq \frac{(b-a)^4}{2^{i+j+2}} M \text{ if } \left| \frac{\partial^2 f(x^*, y^*)}{\partial x \partial y} \right| \leq M$$

Therefore,

$$c_{ij}'^2 = c_{ij}^2 \frac{(b-a)^2}{2^{i+j}} \leq \frac{(b-a)^{10}}{2^{3i+3j+4}} M^2$$

$$\sum_{n=k}^{\infty} \sum_{m=l}^{\infty} c_{nm}'^2 \leq \sum_{n=2^{\alpha+1}}^{\infty} \sum_{m=2^{\beta+1}}^{\infty} \frac{(b-a)^{10}}{2^{3i+3j+4}} M^2, \ \alpha, \beta = 0, 1, 2, \ldots$$

$$\leq (b-a)^{10} M^2 \sum_{n=2^{\alpha+1}}^{\infty} \sum_{i=\beta+1}^{\infty} \sum_{m=2^i}^{2^{i+1}-1} 2^{-3i-3j-4}$$

$$\leq (b-a)^{10} M^2 \sum_{n=2^{\alpha+1}}^{\infty} 2^{-3j-4} \sum_{i=\beta+1}^{\infty} (2^{i+1} - 1 - 2^i + 1) 2^{-3i}$$

$$\leq (b-a)^{10} M^2 \sum_{n=2^{\alpha+1}}^{\infty} 2^{-3j-4} \sum_{i=\beta+1}^{\infty} 2^{-2i}$$

$$\leq (b-a)^{10} M^2 \sum_{n=2^{\alpha+1}}^{\infty} 2^{-3j-4} \, 2^{-2(\beta+1)} \frac{1}{\left[1-(1/2^2)\right]}$$

$$\leq \frac{4(b-a)^{10}}{3l^2} M^2 2^{-4} \sum_{j=\alpha+1}^{\infty} \sum_{n=2^j}^{2^{j+1}-1} 2^{-3j}$$

$$\leq \frac{4(b-a)^{10}}{3l^2} M^2 2^{-4} \sum_{j=\alpha+1}^{\infty} 2^{-2j}$$

$$\leq \frac{4(b-a)^{10}}{3l^2} M^2 2^{-4} \left(\frac{4}{3} \right) 2^{-2(\alpha+1)}$$

$$\leq \left(\frac{16}{9} \right) \frac{(b-a)^{10}}{l^2 k^2} M^2 2^{-4}$$

$$= \left(\frac{16}{144} \right) \frac{(b-a)^{10}}{l^2 k^2} M^2$$

Next,

$$\sum_{n=k}^{\infty}\sum_{m=0}^{l-1}c_{nm}'^{2} \le \sum_{n=k}^{\infty}\sum_{m=0}^{l-1}\frac{(b-a)^{10}M^{2}}{2^{3i+3j+4}} \le \sum_{n=2^{\alpha+1}}^{\infty}\frac{(b-a)^{10}M^{2}}{2^{3j+4}}\sum_{i=0}^{\beta}\sum_{m=2^{i}-1}^{2^{i+1}-1}2^{-3i}$$

$$\le \sum_{n=2^{\alpha+1}}^{\infty}\frac{(b-a)^{10}M^{2}}{2^{3j+4}}\sum_{i=0}^{\beta}\left(2^{-2i}+2^{-3i}\right)$$

$$\le \left(\frac{52}{21}\right)2^{-4}(b-a)^{10}M^{2}\sum_{j=\alpha+1}^{\infty}\sum_{n=2^{j}}^{2^{j+1}-1}2^{-3j}$$

$$\le \left(\frac{52}{336}\right)(b-a)^{10}M^{2}\sum_{j=\alpha+1}^{\infty}2^{-2j}$$

$$\le \left(\frac{52}{336}\right)(b-a)^{10}M^{2}\left\{\frac{2^{-2(\alpha+1)}}{\left[1-(1/2^{2})\right]}\right\}$$

$$= \frac{52(b-a)^{10}M^{2}}{252k^{2}}$$

Similarly, we have

$$\sum_{n=0}^{k-1}\sum_{m=l}^{\infty}c_{nm}'^{2} \le \frac{52(b-a)^{10}M^{2}}{252l^{2}}$$

Then,

$$\sum_{n=k}^{\infty}\sum_{m=l}^{\infty}c_{nm}'^{2} + \sum_{n=k}^{\infty}\sum_{m=0}^{l-1}c_{nm}'^{2} + \sum_{n=0}^{k-1}\sum_{m=l}^{\infty}c_{nm}'^{2} \le \left(\frac{16}{144}\right)\frac{(b-a)^{10}}{l^{2}k^{2}}M^{2}$$

$$+ \frac{52(b-a)^{10}M^{2}}{252k^{2}} + \frac{52(b-a)^{10}M^{2}}{252l^{2}}$$

Hence, we obtain

$$\left\|f(x,y)-f_{kl}(x,y)\right\| \le \frac{(b-a)^{10}M^{2}}{3}\left(\frac{1}{3l^{2}k^{2}}+\frac{13}{21k^{2}}+\frac{13}{21l^{2}}\right)$$

As $l \to \infty$ and $k \to \infty$, we can get $\left\|f(x,y)-f_{kl}(x,y)\right\| \to 0$.

8.10 Conclusion

In this research work, we have successfully applied HWCM to obtain the numerical approximation solution for classical order [110] and fractional order stationary neutron transport equation in a homogeneous isotropic medium. The cited numerical examples well establish that there is a good agreement of results in comparison with the classical solutions. By analyzing Tables 8.1 through 8.12, it can be concluded that if we increase the number of collocation point m, we can get more accurate solution for transport problems. This research work proves the validity and the great potential applicability of the HWCM for solving both the classical and fractional order stationary neutron transport equation.

References

1. Liao, S.J., 2004. On the homotopy analysis method for nonlinear problems, *Applied Mathematics and Computation*, 147, 499–513.
2. Liao, S.J., 2005. Comparison between the homotopy analysis method and homotopy perturbation method, *Applied Mathematics and Computation*, 169, 1186–1194.
3. Adomian, G., 1994. *Solving Frontier Problems of Physics: Decomposition Method*. Kluwer Academic Publishers, Boston, MA.
4. Adomian, G., 1989. *Nonlinear Stochastic Systems Theory and Applications to Physics*. Kluwer Academic Publishers, Dordrecht, the Netherlands.
5. Wazwaz, A., 1999. A reliable modification of Adomian decomposition method, *Applied Mathematics and Computation*, 102(1), 77–86.
6. Cherruault, Y., 1989. Convergence of Adomian's method, *Kybernetes*, 18, 31–38.
7. Cherruault, Y., Adomian, G., 1993. Decomposition methods: A new proof of convergence, *Mathematics and Computer Modeling*, 18, 103–106.
8. He, J.H., 1999. Variational iteration method—A kind of non-linear analytical technique: Some examples, *International Journal of Nonlinear Mechanics*, 34, 699–708.
9. He, J.H., 2000. A review on some new recently developed nonlinear analytical techniques, *International Journal of Nonlinear Sciences and Numerical Simulation*, 1, 51–70.
10. Sweilam, N.H., 2007. Variational iteration method for solving cubic nonlinear Schrodinger equation, *Journal of Computational and Applied Mathematics*, 207, 155–163.
11. Mehdi, T., Mehedi, D., 2007. On the convergence of He's variational iteration method, *Journal of Computational and Applied Mathematics*, 207, 121–128.
12. Smith, G.D., 1985. *Numerical Solution of Partial Differential Equations: Finite Difference Methods* (3rd Edition). Oxford Applied Mathematics and Computing Science Series. Oxford University Press, New York.
13. Yuste, S.B., Acedo, L., 2005. On an explicit finite difference method for fractional diffusion equations, *SIAM Journal on Numerical Analysis*, 52, 1862–1874.
14. Zhou, J.K., 1986. *Differential Transformation and Its Applications for Electrical Circuits*. Huazhong University Press, Wuhan, People's Republic of China.
15. Odibat, Z.M., Bertelle, C., Aziz-Alaoui, M.A., Duchamp, G.H.E., 2010. A multistep differential transform method and application to non-chaotic or chaotic systems, *Computers and Mathematics with Applications*, 59(4), 1462–1472.
16. Arikoglu, A., Ozkol, I., 2005. Solution of boundary value problems for integro-differential equations by using differential transform method, *Applied Mathematics and Computation*, 168(2), 1145–1158.
17. Arikoglu, A., Ozkol, I., 2006. Solution of difference equations by using differential transform method, *Applied Mathematics and Computation*, 174(2), 1216–1228.
18. Liu, H., Song, Y., 2007. Differential transform method applied to high index differential-algebraic equations, *Applied Mathematics and Computation*, 184(2), 748–753.
19. Higham, D.J., 2001. An algorithmic introduction to numerical simulation of stochastic differential equations, *SIAM Review*, 43, 525–546.

20. Sauer, T., 2012. Numerical solution of stochastic differential equation in Finance. In: J.C. Duan, W.K. Härdle, J.E. Gentle (Editors), *Handbook of Computational Finance*. Springer, Berlin, Heidenberg, Part 4, 529–550.

21. Kloeden, P.E., Platen, E., 1992. *Numerical Solution of Stochastic Differential Equations*. Springer-Verlag, New York.

22. Stacey, W.M., 2001. *Nuclear Reactor Physics*. Wiley, Germany.

23. Glasstone, S., Sesonske, A., 1981. *Nuclear Reactor Engineering* (3rd Edition). VNR, New York.

24. Hetrick, D.L., 1993. *Dynamics of Nuclear Reactors*. American Nuclear Society, Chicago, IL.

25. He, J.H., 1997. A new approach to nonlinear partial differential equations, *Communications in Nonlinear Science and Numerical Simulation*, 2, 230–235.

26. Inokti, M., Sekin, H., Mura, T., 1978. General use of the Lagrange multiplier in nonlinear mathematical physics. In: S. Nemat-Nassed (Editor). *Variational Method in the Mechanics of Solids*. Pergamon Press, New York, 156–162.

27. Wazwaz, A.M., El-Sayed, S.M., 2001. A new modification of the Adomian decomposition method for linear and nonlinear operators, *Applied Mathematics and Computation*, 122(3), 393–405.

28. Saha Ray, S., 2008. An application of the modified decomposition method for the solution of the coupled Klein–Gordon–Schrödinger equation, *Communication in Nonlinear Science and Numerical Simulation*, 13, 1311–1317.

29. Kaya, D., Yokus, A., 2002. A numerical comparison of partial solutions in the decomposition method for linear and nonlinear partial differential equations, *Mathematics and Computers in Simulation*, 60(6), 507–512.

30. Büyük, S., Çavdar, S., Ahmetolan, S., 2010. Solution of fixed source neutron diffusion equation via homotopy perturbation method. In: *ICAST 2010*, May, İzmir, Turkey (in English).

31. Saha Ray, S., Patra, A., 2011. Application of modified decomposition method and variational iteration method for the solution of the one group neutron diffusion equation with fixed source, *International Journal of Nuclear Energy Science and Technology*, 6(4), 310–320.

32. Duderstadt, J.J., Hamilton, L.J., 1976. *Nuclear Reactor Analysis*. Wiley, New York.

33. Liao, S.J., 1997. Homotopy analysis method: A new analytical technique for nonlinear problems, *Communications in Nonlinear Science and Numerical Simulation*, 2(2), 95–100.

34. Liao, S.J., 2003. *Beyond Perturbation: Introduction to the Homotopy Analysis Method*. CRC Press, Boca Raton, FL.

35. Saha Ray, S., Patra, A., 2011. Application of homotopy analysis method and Adomian decomposition method for the solution of neutron diffusion equation in the hemisphere and cylindrical reactors, *Journal of Nuclear Engineering and Technology*, 1(1–3), 1–14.

36. Cassell, J.S., Williams, M.M.R., 2004. A solution of the neutron diffusion equation for a hemisphere with mixed boundary conditions, *Annals of Nuclear Energy*, 31, 1987–2004.

37. Williams, M.M.R., Matthew, E., 2004. A solution of the neutron diffusion equation for a hemisphere containing a uniform source, *Annals of Nuclear Energy*, 31, 2169–2184.

38. Adomian, G., 1991. A review of the decomposition method and some recent results for nonlinear equations, *Computers & Mathematics with Applications*, 21, 101–127.

39. Dababneh, S., Khasawneh, K., Odibat, Z., 2011. An alternative solution of the neutron diffusion equation in cylindrical symmetry, *Annals of Nuclear Energy*, 38, 1140–1143.

40. Khasawneh, K., Dababneh, S., Odibat, Z., 2009. A solution of the neutron diffusion equation in hemispherical symmetry using the homotopy perturbation method, *Annals of Nuclear Energy*, 36, 1711–1717.

41. Kilbas, A.A., Srivastava, H.M., Trujillo, J.J., 2006. *Theory and Applications of Fractional Differential Equations*. North-Holland Mathematical Studies, 204. Elsevier (North-Holland) Science Publishers, Amsterdam, the Netherlands.

42. Sabatier, J., Agrawal, O.P., Tenreiro Machado, J.A., 2007. *Advances in Fractional Calculus: Theoretical Developments and Applications in Physics and Engineering*. Springer, Dordrecht, the Netherlands.

43. Gorenflo, R., Mainardi, F., 1997. Fractional calculus integral and differential equations of fractional order. In: A. Carpinteri, F. Mainardi (Editors), *Fractals and Fractional Calculus in Continuum Mechanics*. Springer-Verlag, New York, 223–276.

44. Podlubny, I., 1999. *Fractional Differential Equations*. Academic Press, San Diego, CA.

45. Holmgren, H.J., 1865–1866. Om differentialkalkylen med indecies of hvad natur som helst, *Kungliga Svenska Vetenskaps-Akademins Handlingar*, 5(11), 1–83, Stockholm (in Swedish).

46. Riemann, B., 1876. Versuch einer allgemeinen Auffassung der Integration und Differentiation. In: *Gesammelte Werke und Wissenschaftlicher Nachlass*. Teubner, Leipzig, Germany, 331–344.

47. Zhang, D.L., Qiu, S.Z., Su, G.H., Liu, C.L., Qian, L.B., 2009. Analysis on the neutron kinetics for a molten salt reactor, *Progress in Nuclear Energy*, 51, 624–636.

48. Espinosa-Paredes, G., Morales-Sandoval, J.B., Vázquez-Rodríguez, R., Espinosa-Martínez, E.-G., 2008. Constitutive laws for the neutron density current, *Annals of Nuclear Energy*, 35, 1963–1967.

49. Espinosa-Paredes, G., Polo-Labarrios, M.-A., Espinosa-Martínez, E.-G., Valle-Gallegos, E.D., 2011. Fractional neutron point kinetics equations for nuclear reactor dynamics, *Annals of Nuclear Energy*, 38, 307–330.

50. Saha Ray, S., Patra, A., 2012. An explicit finite difference scheme for numerical solution of fractional neutron point kinetic equation, *Annals of Nuclear Energy*, 41, 61–66.

51. Lamarsh, J.R., Baratta, A.J., 2001. *Introduction to Nuclear Engineering* (3rd Edition). Prentice Hall, Upper Saddle River, NJ.

52. Sardar, T., Saha Ray, S., Bera, R.K., Biswas, B.B., 2009. The analytical approximate solution of the multi-term fractionally damped Vander Pol equation, *Physica Scripta*, 80, 025003.

53. Kinard, M., Allen, K.E.J., 2004. Efficient numerical solution of the point kinetics equations in nuclear reactor dynamics, *Annals of Nuclear Energy*, 31, 1039–1051.

54. Edwards, J.T., Ford, N.J., Simpson, A.C., 2002. The numerical solution of linear multiterm fractional differential equations, *Journal of Computational and Applied Mathematics*, 148, 401–418.

55. Diethelm, K., 1997. An algorithm for the numerical solution of differential equations of fractional order, *Electronic Transactions on Numerical Analysis*, 5, 1–6.

56. Diethelm, K., 1997. Numerical approximation of finite-part integrals with generalized compound quadrature formulae, *IMA Journal of Numerical Analysis*, 17, 479–493.

57. Khan, Y., Faraz, N., Yildirim, A., Wu, Q., 2011. Fractional variational iteration method for fractional initial-boundary value problems arising in the application of nonlinear science, *Computers & Mathematics with Applications*, 62(5), 2273–2278.

58. Khan, Y., Wu, Q., Faraz, N., Yildirim, A., Madani, M., 2012. A new fractional analytical approach via a modified Riemann-Liouville derivative, *Applied Mathematics Letters*, 25(10), 1340–1346.

59. Khan, Y., Fardi, M., Seyevand, K., Ghasemi, M., 2014. Solution of nonlinear fractional differential equations using an efficient approach, *Neural Computing and Applications*, 24, 187–192.

60. Hayes, J.G., Allen, E.J., 2005. Stochastic point-kinetics equations in nuclear reactor dynamics, *Annals of Nuclear Energy*, 32, 572–587.

61. Darania, P., Ebadian, A., 2007. A method for the numerical solution of the integro-differential equations, *Applied Mathematics and Computation*, 188, 657–668.

62. Patrico, M.F., Rosa, P.M., 2007. The differential transform method for advection-diffusion problems, *International Journal of Mathematical, Computational Science and Engineering*, 1(9), 73–77.

63. Quintero-Leyva, B., 2008. CORE: A numerical algorithm to solve the point kinetics equations with lumped model temperature and feedback, *Annals of Nuclear Energy*, 36, 246–250.

64. McMohan, D., Pierson, A., 2010. A Taylor series solution of the reactor point kinetic equations, http://arxiv.org/ftp/arxiv/papers/1001/1001.4100.pdf.

65. Van den Eynde, G., 2006. A resolution of the stiffness problem of reactor kinetics, *Nuclear Science and Engineering*, 153, 200–202.

66. Patra, A., Saha Ray, S., 2013. Multi-step differential transform method for numerical solution of classical neutron point kinetic equation, *Computational Mathematics and Modeling*, 24(4), 604–615.

67. Saha Ray, S., Patra, A., 2014. Numerical simulation for solving fractional neutron point kinetic equations using the multi-step differential transform method, *Physica Scripta*, 89, 015204.

68. Espinosa-Paredes, G., Polo-Labarrios, M.-A., Diaz-Gonzalez, L., Vazquez-Rodriguez, A., Espinosa-Matinez, E.-G., 2012. Sensitivity and uncertainty analysis of the fractional neutron point kinetics equations, *Annals of Nuclear Energy*, 42, 169–174.

69. Polo-Labarrios, M.-A., Espinosa-Paredes, G., 2012. Application of the fractional neutron point kinetic equation: Start-up of a nuclear reactor, *Annals of Nuclear Energy*, 46, 37–42.

70. Polo-Labarrios, M.-A., Espinosa-Paredes, G., 2012. Numerical analysis of start-up PWR with fractional neutron point kinetic equation, *Progress in Nuclear Energy*, 60, 38–46.

71. Saha Ray, S., 2012. Numerical simulation of stochastic point kinetic equation in the dynamical system of nuclear reactor, *Annals of Nuclear Energy*, 49, 154–159.

72. Oldham, K.B., Spanier, J., 1974. *The Fractional Calculus*. Academic Press, New York.

73. Saha Ray, S., Patra, A., 2012. Numerical solution for stochastic point kinetics equations with sinusoidal reactivity in dynamical system of nuclear reactor, *International Journal of Nuclear Energy Science and Technology*, 7(3), 231–242.

74. Patra, A., Saha Ray, S., 2014. The effect of pulse reactivity for stochastic neutron point kinetics equation in nuclear reactor dynamics, *International Journal of Nuclear Energy Science and Technology*, 8(2), 117–130.

75. Saha Ray, S., Patra, A., 2013. Numerical solution of fractional stochastic neutron point kinetic equation for nuclear reactor dynamics, *Annals of Nuclear Energy*, 54, 154–161.

76. Nahla, A.A., 2008. Generalization of the analytical exponential model to solve the point kinetics equations of Be- and D_2O-moderated reactors, *Nuclear Engineering and Design*, 238, 2648–2653.

77. Aboander, A.E., Nahla, A.A., 2002. Solution of the point kinetics equations in the presence of Newtonian temperature feedback by Padé approximations via the analytical inversion method, *Journal of Physics A: Mathematical and General*, 35, 9609–9627.

78. Nahla, A.A., 2010. Analytical solution to solve the point reactor kinetics equations, *Nuclear Engineering and Design*, 240, 1622–1629.

79. Nahla, A.A., 2011. Taylor's series method for solving the nonlinear point kinetics equations, *Nuclear Engineering and Design*, 241, 1592–1595.

80. Li, H., Chen, W., Luo, L., Zhu, Q., 2009. A new integral method for solving the point reactor neutron kinetics equations, *Annals of Nuclear Energy*, 36, 427–432.

81. Chao, Y., Attard, A., 1985. A resolution of the stiffness problem of reactor kinetics, *Nuclear Science and Engineering*, 90, 40–46.

82. Miller, K.S., Ross, B., 1993. *An Introduction to the Fractional Calculus and Fractional Differential Equations*. Wiley, New York.

83. Patra, A., Saha Ray, S., 2014. On the solution of nonlinear fractional neutron point kinetics equation with Newtonian temperature feedback reactivity, *Nuclear Technology*, 189, 103–109.

84. Nahla, A.A., 2011. An efficient technique for the point kinetics equations with Newtonian temperature feedback effects, *Annals of Nuclear Energy*, 38, 2810–2817.

85. Aboanber, A.E., Nahla, A.A., 2003. Analytical solution of the point kinetics equations by exponential mode analysis, *Progress in Nuclear Energy*, 42, 179–197.

86. Aboanber, A.E., Nahla, A.A., 2002. Generalization of the analytical inversion method for the solution of the point kinetics equations, *Journal of Physics A: Mathematical and General*, 35, 3245–3263.

87. Hamada, Y.M., 2011. Generalized power series method with step size control for neutron kinetics equations, *Nuclear Engineering and Design*, 241, 3032–3041.

88. Ganapol, B.D., 2013. A highly accurate algorithm for the solution of the point kinetics equations, *Annals of Nuclear Energy*, 62, 564–571.

89. Picca, P., Furfaro, R., Ganapol, B.D., 2013. A highly accurate technique for the non-linear point kinetics equations, *Annals of Nuclear Energy*, 58, 43–53.

90. Chui, C.K., 1992. *An Introduction to Wavelets*. Academic Press, London.

91. Mallat, S.G., 2009. *A Wavelet Tour of Signal Processing: The Sparse Way* (3rd Edition). Academic Press, San Diego, CA.

92. Debnath, L., 2007. *Wavelet Transforms and Their Applications*. Birkhäuser, Boston, MA.

93. Haar, A., 1910. Zur theorie der orthogonalen Funktionsysteme, *Mathematical Analysis*, 69, 331–371.

94. Li, Y., Zhao, W., 2010. Haar wavelet operational matrix of fractional order integration and its applications in solving the fractional order differential equations, *Applied Mathematics and Computation*, 216, 2276–2285.

95. Saha Ray, S., 2012. On Haar wavelet operational matrix of general order and its application for the numerical solution of fractional Bagley Torvik equation, *Applied Mathematics and Computation*, 218, 5239–5248.

194   *References*

96. Chen, C.F., Hsiao, C.H., 1997. Haar wavelet method for solving lumped and distributed parameter-systems, *IEE Proceedings—Control Theory and Applications*, 144(1), 87–94.
97. Saha Ray, S., Patra, A., 2013. Haar wavelet operational methods for the numerical solutions of fractional order nonlinear oscillatory Van der Pol system, *Applied Mathematics and Computation*, 220, 659–667.
98. Ganapol, B.D., Picca, P., Previti, A., Mostacci, D., 2012. *The Solution of the Point Kinetics Equations via Convergence Acceleration Taylor Series (CATS)*. Knoxville, TN.
99. Baleanu, D., Kadem, A., 2011. About the F_N Approximation to fractional neutron transport equation in slab geometry. In: *Proceedings of the ASME International Design Engineering Technical Conference and Computers and Information in Engineering Conference*, Washington, DC.
100. Kadem, A., Kilicman, A., 2011. Note on transport equation and fractional Sumudu transform, *Computers and Mathematics with Applications*, 62, 2995–3003.
101. Kadem, A., Baleanu, D., 2011. Solution of a fractional transport equation by using the generalized quadratic form, *Communications in Nonlinear Science and Numerical Simulation*, 16, 3011–3014.
102. Kulikowska, T., 2000. An introduction to the neutron transport phenomena. In: *Proceedings of the Workshop on Nuclear Data and Nuclear Reactors: Physics, Design and Safety*, Trieste, Italy.
103. Martin, O., 2011. A homotopy perturbation method for solving a neutron transport equation, *Applied Mathematics and Computation*, 217, 8567–8574.
104. Šmarda, Z., Khan, Y., 2012. Singular initial value problem for a system of integro-differential equations, *Abstract and Applied Analysis*, 2012, Article Id 918281.
105. Islam, S., Imran, A., Muhammad, F., 2013. A new approach for numerical solution of integro-differential equations via Haar wavelets, *International Journal of Computer Mathematics*, 90, 1971–1989.
106. Chang, P., Piau, P., 2008. Haar wavelet matrices designation in numerical solution of ordinary differential equations, *IAENG International Journal of Applied Mathematics*, 38, 3–11.
107. Chen, Y., Wu, Y., Cui, Y., Wang, Z., Jin, D., 2010. Wavelet method for a class of fractional convection-diffusion equation with variable coefficients, *Journal of Computational Science*, 1, 146–149.
108. Lepik, U., 2009. Solving fractional integral equations by the Haar wavelet method, *Applied Mathematics and Computation*, 214, 468–478.
109. Martin, O., 2006. Numerical algorithm for a transport equation with periodic source function, *International Journal of Information and Systems Sciences*, 2, 436–451.
110. Patra, A., Saha Ray, S., 2014. Two-dimensional Haar wavelet collocation method for the solution of stationary neutron transport equation in a homogeneous isotropic medium, *Annals of Nuclear Energy*, 70, 30–35.

Index

Note: Locators "*f*" and "*t*" denote figures and tables in the text